2013
万达商业规划
销售类物业

WANDA COMMERCIAL
PLANNING 2013
PROPERTIES FOR SALE

万达商业地产设计中心 主编

中国建筑工业出版社

EDITORIAL BOARD MEMBERS
编委会成员

主编单位
万达商业地产设计中心

规划总指导
丁本锡　齐界

编委
赖建燕　曲晓东　于修阳　吕正韬　尹强　林树郁
刘拥军　曾静　张东光　王福魁

参编人员
白雪松　付东滨　孙志超　刘大伟　李金桥　昌燕　黄建好
俞小华　薛勇　刘征　文善平

郑锐　周昳晗　牛辉哲　李春阳　单萍　张舰文　周恒
武春雨

赵龙　赵立群　李军　赵宁宁　杜文天　靳松　李晅荣　曹鹏
叶啸　孙静　董莉　吴铮　陈文娜　黄达志　杨威　钟山　杨蓦
刘洋　杨磊　宋之煜　李琰　李万顺　胡延峰　杨琼　范志满
张金菊　朱卓毅　冯俊　董丽梅　朱应勇　赵前卫　陈涛
桑国安　彭亚飞　罗粤峰　邹昊　邵强　陈海燕　顾东方　李鹏
汪新华　杜刘生　薛奇　苏嘉峰　关向东　胡楠　霍雪影　刘国祥
宋波

校对
白雪松　郑锐　刘大伟　孙志超　昌燕　李金桥　俞小华
黄建好　袁文卿

CHIEF EDITORIAL UNIT
Wanda Commercial Estate Design Center

GENERAL PLANNING DIRECTORS
Ding Benxi, Qi Jie

EDITORIAL BOARD
Lai Jianyan, Qu Xiaodong, Yu Xiuyang, Lv Zhengtao, Yin Qiang, Lin Shuyu, Liu Yongjun, Zeng Jing, Zhang Dongguang, Wang Fukui

PARTICIPANTS
Bai Xuesong, Fu Dongbin, Sun Zhichao, Liu Dawei, Li Jinqiao, Chang Yan, Huang Jianhao, Yu Xiaohua, Xue Yong, Liu Zheng, Wen Shanping

Zheng Rui, Zhou Yihan, Niu Huizhe, Li Chunyang, Shan Ping, Zhang Jianwen, Zhou Heng, Wu Chunyu

Zhao Long, Zhao Liqun, Li Jun, Zhao Ningning, Du Wentian, Jin Song, Li Xuanrong, Cao Peng, Ye Xiao, Sun Jing, Dong Li, Wu Zheng, Chen Wenna, Huang Dazhi, Yang Wei, Zhong Shan, Yang Mo, Liu Yang, Yang Lei, Song Zhiyu, Li Yan, Li Wanshun, Hu Yanfeng, Yang Qiong, Fan Zhiman, Zhang Jinju, Zhu Zhuoyi, Feng Jun, Dong Limei, Zhu Yingyong, Zhao Qianwei, Chen Tao, Sang Guoan, Peng Yafei, Luo Yuefeng, Zou Hao, Shao Qiang, Chen Haiyan, Gu Dongfang, Li Peng, Wang Xinhua, Du Liusheng, Xue Qi, Su Jiafeng, Guan Xiangdong, Hu Nan, Huo Xueying, Liu Guoxiang, Song Bo

PROOFREADERS
Bai Xuesong, Zheng Rui, Liu Dawei, Sun Zhichao, Chang Yan, Li Jinqiao, Yu Xiaohua, Huang Jianhao, Yuan Wenqing

CONTENTS
目录

PART A FOREWORD
序言 014

016 / **WANDA COMMERCIAL ESTATE DESIGN CENTER AND WANDA PROPERTIES FOR SALE**
万达商业地产设计中心与万达销售物业

PART B WANDA MANSIONS & WANDA PALACES
万达公馆、华府 018

020 / **DESIGN CONTROL FEATURES OF WANDA MANSIONS & WANDA PALACES**
万达公馆、华府设计管控特点

WANDA MANSIONS
公馆系列

024 / **DALIAN DONGGANG WANDA MANSION**
大连东港万达公馆

028 / **FUJIAN QUANZHOU WANDA MANSION**
福建泉州万达公馆

036 / **CHANGSHA KAIFU WANDA MANSION**
长沙开福万达公馆

048 / **WUHAN JIYUQIAO WANDA MANSION**
武汉积玉桥万达公馆

054 / **JINAN WEIJIAZHUANG WANDA MANSION**
济南魏家庄万达公馆

060 / **SHIJIAZHUANG YUHUA WANDA MANSION**
石家庄裕华万达公馆

WANDA PALACES
华府系列

066 / **ZHANGZHOU BIHU WANDA PALACE**
漳州碧湖万达华府

070 / **ZHENGZHOU ERQI WANDA PALACE**
郑州二七万达华府

074 / **FUJIAN PUTIAN WANDA PALACE**
福建莆田万达华府

PART C **WANDA SALES PLACES**
万达销售卖场
078

080 / **DESIGN TRILOGY OF WANDA SALES PLACES**
万达销售卖场设计三部曲

SALES PLACE OF WANDA MANSIONS
公馆卖场

086 / **SHENYANG OLYMPIC WANDA MANSION: SALES OFFICE/ PROTOTYPE ROOM**
沈阳奥体万达公馆－售楼处／样板间

094 / **NANNING QINGXIU WANDA MANSION: SALES OFFICE/ PROTOTYPE ROOM**
南宁青秀万达公馆－售楼处／样板间

098 / **CHANGSHA KAIFU WANDA MANSION: SALES OFFICE/ PROTOTYPE ROOM**
长沙开福万达公馆－售楼处／样板间

104 / **DONGGUAN DONGCHENG WANDA MANSION: SALES OFFICE/ PROTOTYPE ROOM**
东莞东城万达公馆－售楼处／样板间

110 / **YANTAI ZHIFU WANDA MANSION: SALES OFFICE/ PROTOTYPE ROOM**
烟台芝罘万达公馆－售楼处／样板间

122 / **NANJING JIANGNING WANDA MANSION: SALES OFFICE**
南京江宁万达公馆－售楼处

124 / **WUHAN K4 WANDA MANSION: SALES OFFICE/ PROTOTYPE ROOM**
武汉 K4 万达公馆－售楼处／样板间

130 / **XI'AN DAMINGGONG WANDA MANSION: SALES OFFICE/ PROTOTYPE ROOM**
西安大明宫万达公馆－售楼处／样板间

136 / **ZHENGZHOU JINSHUI WANDA MANSION: SALES OFFICE**
郑州金水万达公馆－售楼处

SALES PLACE OF WANDA PALACES
华府卖场

140 / **DONGGUAN HOUJIE WANDA PALACE: SALES OFFICE**
东莞厚街万达华府－售楼处

146 / **DEZHOU WANDA PALACE: SALES OFFICE/ PROTOTYPE ROOM**
德州万达华府－售楼处／样板间

152 / **ZHEJIANG JIAXING WANDA PALACE: PROTOTYPE ROOM**
浙江嘉兴万达华府－样板间

156 / **JIAMUSI WANDA PALACE: SALES OFFICE**
佳木斯万达华府－售楼处

160 / **MIANYANG CBD WANDA PALACE: PROTOTYPE ROOM**
绵阳CBD万达华府－样板间

164 / **WEINAN WANDA PALACE: SALES OFFICE/ PROTOTYPE ROOM**
渭南万达华府－售楼处／样板间

170 / **FOSHAN NANHAI WANDA PALACE: PROTOTYPE ROOM**
佛山南海万达华府－样板间

174 / **JINING WANDA PALACE: PROTOTYPE ROOM**
济宁万达华府－样板间

178 / **WENZHOU PINGYANG WANDA PALACE: PROTOTYPE ROOM**
温州平阳万达华府－样板间

180 / **CHONGQING BA'NAN WANDA PALACE: SALES OFFICE**
重庆巴南万达华府－售楼处

PART D / THE DEMONSTRATION AREAS OF WANDA SALES PLACES
万达销售实景示范区

184 / **STRATEGY OF THE DEMONSTRATION AREA OF WANDA SALES PLACES**
万达销售实景示范区展示策略

THE DEMONSTRATION AREAS OF WANDA MANSIONS
公馆示范区

190 / **DEMONSTRATION AREA OF DALIAN HIGH-TECH ZONE WANDA MANSION**
大连高新万达公馆示范区

198 / **DEMONSTRATION AREA OF DONGGUAN DONGCHENG WANDA MANSION**
东莞东城万达公馆示范区

206 / **DEMONSTRATION AREA OF TIANJIN HAIHEDONGLU COMMERCIAL STREET**
天津海河东路商业街示范区

212 / **DEMONSTRATION AREA OF NANNING QINGXIU WANDA MANSION**
南宁青秀万达公馆示范区

216 / **DEMONSTRATION AREA OF WUHAN K9 WANDA MANSION**
武汉K9万达公馆示范区

222 / **DEMONSTRATION AREA OF WUXI YIXING WANDA MANSION**
无锡宜兴万达公馆示范区

228 / **DEMONSTRATION AREA OF XI'AN DAMINGGONG WANDA MANSION**
西安大明宫万达公馆示范区

THE DEMONSTRATION AREAS OF WANDA PALACES
华府示范区

232 / **DEMONSTRATION AREA OF MA'ANSHAN WANDA PALACE**
马鞍山万达华府示范区

238 / **DEMONSTRATION AREA OF QINGDAO LICANG WANDA PALACE COMMERCIAL STREET**
青岛李沧万达华府
商业街示范区

242 / **DEMONSTRATION AREA OF DONGGUAN CHANG'AN WANDA PALACE**
东莞长安万达华府示范区

244 / **DEMONSTRATION AREA OF FOSHAN NANHAI WANDA PALACE**
佛山南海万达华府示范区

248 / **DEMONSTRATION AREA OF CHANGZHOU WUJIN WANDA PALACE**
常州武进万达华府示范区

256 / **DEMONSTRATION AREA OF WEIFANG WANDA PALACE**
潍坊万达华府示范区

260 / **DEMONSTRATION AREA OF ANHUI BENGBU WANDA PALACE COMMERCIAL STREET**
安徽蚌埠万达华府商业街示范区

PART E CLASSIC PROJECTS RETROSPECT OF WANDA PROPERTIES FOR SALE
万达销售物业经典项目回顾

262

264 / **REPRODUCTION OF TRADITION:**
RETROSPECT OF CLASSIC PROJECTS OF THE CHINESE-STYLE COURTYARDS OF WANDA ORGANIC AGRICULTURE GARDEN
古为今用再造传统——万达有机农业园中式四合院经典项目回顾

268 / **WANDA CHINESE-STYLE COURTYARD**
万达中式四合院

PART F / EPILOGUE
后续 — 278

280 / RETROSPECT OF WANDA PROPERTIES FOR SALE
万达销售物业历程回顾

282 / RELATIONSHIP BETWEEN WANDA PROPERTIES FOR SALE AND WANDA PLAZAS: "AGGREGATED VALUE" IN THE RESOURCE SYMBIOSIS ERA
万达销售物业与万达广场的关系
——资源共生时代下的"聚合增值"

286 / WANDA PROPERTIES FOR SALE 2013
万达销售物业 2013

290 / POSTSCRIPT
后记

PART G / PROJECT INDEX
项目索引 — 292

294 / ADMISSION PROJECTS
入伙项目

295 / SALES PLACES
销售卖场

297 / DEMONSTRATION AREAS
示范区

PART **A** WANDA COMMERCIAL PLANNING
FOREWORD
序言

WANDA COMMERCIAL ESTATE DESIGN CENTER AND WANDA PROPERTIES FOR SALE

万达商业地产设计中心与万达销售物业

文/万达商业地产设计中心总经理 尹强

万达销售物业产品设计是万达商业地产成功模式的重要组成部分。

万达广场产品系列是万达独创的商业地产产品,包括大型商业中心、室外商业街、五星级酒店、写字楼和公寓住宅等;其销售物业产品类型主要包含住宅、城市商业街、写字楼和公寓四种核心产品,以及销售商务酒店、销售酒楼、车位等其他产品(图1)。

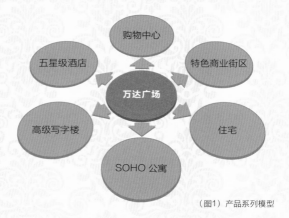

(图1)产品系列模型

25年来,随着万达集团的发展,销售物业产品品质大幅度提升,为业内所瞩目。

万达商业地产设计中心是万达集团总部销售物业产品的设计管控部门。从2008年至今,为了确保产品的品质、设计进度、安全、质量等核心目标,并满足开发现金流的需求,设计中心形成了以设计管理为基础、以满足并引领客户需求为目标,不断总结创新的销售物业设计管理体系。

设计中心从2008年的4个人,发展为2013年的110人,管控范围历经变迁,逐年深化。伴随着万达商业地产的不断发展,商业地产设计中心对销售物业产品管控的深度从单一建筑专业到全专业,从方案到施工图以及现场巡查,从住宅到销售物业全部产品类型,从管控国内项目到管控境外项目。

为了加强销售类物业的管控,2009年万达商业地产成立了项目管理中心设计部,编制10人;2010年,扩编为18人;2011年,项目管理中心设计部分为南、北方项目管理中心设计部,编制共36人;2012年,再次扩编为90人,对销售物业的设计管控扩展到了施工图设计和现场设计配合,实现了对销售物业的全

The product design of Wanda Properties for Sale is an important component of the successful pattern of Wanda Commercial Properties.

The product series of Wanda Plazas are unique commercial real estate products created by Wanda, including large-scale business center, outdoor commercial pedestrian streets, five-star hotels, office buildings, and apartments, etc. The types of the Properties for Sale mainly consist of four core products, which are residences, commercial pedestrian streets, office buildings and apartments, and other products including sales business hotels, sales restaurants, and parking lots, etc (Figure 1).

In the past twenty five years, with the development of Wanda Group, the product quality of Wanda Properties for Sale have been greatly improved and caught the attention of the industry.

Wanda Commercial Estate Design Center is the design control department of Wanda Properties for Sale. Since 2008, in order to ensure the achievements of our main goals, such as product quality, design progress, safety, etc., and to satisfy the needs for cash flow, the Design Center has formed a new design management system of Properties for Sale which is based on design management, targets to satisfy and lead clients' demands and continuously realizes innovation.

The Design Center has increased from 4 to 110 employees from 2008 to 2013. Along with the constant development of Wanda Commercial Properties, the Design Center expands the design control scopes of Properties for Sale year after year, from architecture to all other specialties, from conception design to construction design and on-site inspection, from residences to all types of Properties for Sale, and from domestic projects to oversea projects.

To strengthen the management of Properties for Sale, Wanda Commercial Estate Co., Ltd. established Design Department in the Project Management Center in 2009 with 10 employees; and it increased to 18 employees in the following year. In 2011, the Design Department is divided into two design departments and both the South and North Project Management Centers had 36 employees. In 2012, it enlarged to 90 employees and the management scope expanded to construction drawing design, coordinating with on-site design, and it has fulfilled the target of management of Properties for Sale during the whole process with all specialties. In 2013, the design departments of the South and North Project Management Centers merged with Wanda Commercial Planning & Research Institute as Wanda Commercial Properties Planning System in 2013 with 110 employees in total.

In 2014, the South and North Project Management Centers merged into Wanda Commercial Estate Design Center,

专业全过程管控；2013年，南、北方项目管理中心设计部与万达商业规划研究院合并为商业地产规划系统，编制增加为110人。

2014年，南、北方项目管理中心设计部合并为商业地产设计中心。设计中心的职能涵盖了商业地产销售类物业的全专业、全阶段设计管理。

设计中心负责制定销售物业设计管理制度。该制度明确了设计基本要求和强制性标准，是设计方、项目公司、集团总部统一的技术标准。设计计划管理措施，是确保管控销售物业各项设计节点按计划保质保量完成的基础。

确保销售物业的设计安全，是万达销售物业设计的重中之重，也是设计中心的重点工作职责。销售物业设计的安全管理体系，划定了不可逾越的安全红线，为万达商业地产顺利发展保驾护航。

销售类物业技术研究和产品研发是设计中心的重要职能。设计决定市场导向，创造市场价值。万达销售物业的高端品质依赖于强大的产品研发，体现了设计中心的专业价值（图2~图4）。

销售卖场的设计全程把控，提前给予客户完美的产品体验；入伙交楼时高品质呈现的产品，是万达商业地产向客户提交的满意答卷，也是设计中心的智慧结晶。

2014年是万达商业地产设计中心的创新年。其管理创新主要体现在以下5个方面：（1）项目管控方面，突出重点，管理创新；（2）在课题研发方面，策划前置，品质优先；（3）在业务培训方面，扩大范围，形式创新；（4）在考核评估方面，细化考核内容，完善考核标准；（5）在产品创新方面，追求卓越的设计，强化方案的管控，体现商业地产开发中的设计价值。

万达集团致力于为消费者提供高品质的产品，对产品品质提升有着不懈的追求。未来几年，万达商业地产规划设计系统将进一步整合资源，提升管控能力。商业地产设计中心将提供更优秀的产品，发挥更大的作用，在产品标准化的基础上实现产品创新、管理创新，走在行业发展的前列。

（图2）长沙开福万达公馆入口大门

（图3）大连东港万达公馆旋转楼梯

（图4）长沙开福公馆活动场地景观绿化组团

whose responsibilities covering design management of full process in all specialties of commercial real estate properties for sale.

The Design Center is in charge of formulating design management regulations of Properties for Sale. It clarifies that the basic requirements of design and mandatory standards are the unified technical standards of design producers, project companies and Group Headquarters. The design planning management is the foundation to guarantee accomplishments with satisfied quality and on schedule in the process of Properties for Sale management.

To ensure safety of design is the priority issue in the design of Wanda Properties for Sale, and is also a significant responsibility of the Design Center. The safety management system of Properties for Sale explicitly demarcates safety red line in order to achieve successful development of Wanda Commercial Properties.

Another important responsibility of the Design Center is Properties for Sale R&D in technical and products. Design determines market orientation and creates market value. The high quality of Wanda Properties for Sale relies on powerful product R&D, reflecting professional values of the Design Center. (Figure 2 to Figure 4).

The Design Center manages Sales Place design in full process to give clients perfect experience in advance and provide high quality products when the buyers moved in. These are the fulfillment of the commitment Wanda Commercial Properties makes to clients, and also are the wisdom crystallization of the Design Center as well.

2014 is a year of innovation for Wanda Commercial Estate Design Center. Its management innovation mainly manifests on the following five aspects: (1) Project management: highlighting key points and management innovation; (2) R&D: advanced planning and quality priority; (3) Business training: extending scopes and innovation; (4) Examination and evaluation: detailing examination contents and maturing assessment criteria; (5) Product innovation: pursuing distinguished design, enhancing design management to reflect design value in the commercial properties development.

Wanda Group is committed to providing high quality products for clients and persistently pursues for product quality improvement. In the coming years, Wanda Commercial Properties Planning and Design System will further integrate with all kinds of resources to improve control and management skills. And the Design Center will create more excellent products and fulfill targets of product innovation and management innovation on the basis of product standardization to play a more important leading role in the development of the industry.

审图号：GS（2014）2552号

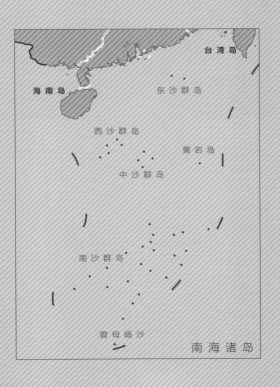

DESIGN CONTROL FEATURES OF WANDA MANSIONS & WANDA PALACES

万达公馆、华府设计管控特点

文/万达商业地产设计中心北区设计部总经理 曾静

万达广场产品业态丰富，各业态间的功能优化组合。一座万达广场将零售、文化、餐饮、娱乐、休闲及酒店等诸多功能整合在一起，不仅一站式地满足并丰富了城市居民的消费需求，还对城市的商业设施进行了完善升级。

销售物业作为万达广场项目的重要组成部分，与持有物业组合，形成从居住、生活、工作到消费娱乐等完整的城市功能。与此同时，万达广场为居住工作人群所提供的完善配套及服务，也正体现了万达销售物业的核心价值。所以，如何充分利用持有物业完善的配套，为客户提供更多附加价值；如何在项目的快速发展中保证销售物业的品质，是万达销售物业管控的难点（图1），下面就公馆和华府两种主要销售物业产品进行阐述。

（图1）武汉积玉桥万达公馆建筑外立面

针对万达公馆、华府特点，万达商业地产设计中心制定了相应的设计管控办法，具体如下。

1. 周密策划，提前启动

在项目挂牌前，即配合营销进行项目市场调研及当地规划条件调查，设定项目的价值点、决策点、控制点，形成详细的可行性研究报告；识别并评估项目风险点，制定预案；同时由代管项目公司启动设计招标，确定设计单位。因此，须在项目摘牌前取得设计启动的必备条件，为项目设计快速推动打下良好的基础。

Wanda Plaza has optimized functional combination of multiple industries. A Wanda Plaza, integrating functions of retail, culture, dining, entertainment, recreation, hotel and so forth, brings urban residents one-stop consumption experience and also updates urban commercial facilities.

As an integral part of Wanda Plaza projects, the Properties for Sale, combining with the Properties for Holding, have formed complete urban functions including residence, life, work, consumption and entertainment. Meanwhile, the well-equipped supporting facilities and services provided by Wanda Plazas for people who are living and working there reflect the core values of Wanda Properties for Sale. So, the difficulties in Wanda Properties for Sale management are how to make full use of the well-equipped supporting facilities of the Properties for Holding to provide more added values for clients, and how to guarantee quality of properties for sale in the rapid project development (Figure 1).

Targeting at the characteristics of Wanda Properties for Sale, Wanda Commercial Estate Design Center has formulated the following design and control methods.

1. WELL PLANNED AND PREPARED FOR EARLY LAUNCH

Before a project is listed, market survey and investigation for local planning conditions should be made combined with marketing activities to determine the value point, decision point and control point of the project. A detailed feasibility study report is produced to identify and assess project risks, and create risk management plan. At the same time, the project escrow company launches bid invitation to select a design firm. Therefore, necessary conditions for design launching must be possessed before the project is delisted in order to lay good foundation for successfully implementation of the design works.

2. FULL PROCESS CONTROL WITH HIGHLIGHTS OF KEY POINTS

Design management of Wanda Properties for Sale is fully professional and full-process management including product positioning, project designing, project refining, construction drawing design and confirmation at the project initial & planning stage and project implementation as well. There are detailed management methods for each specialty in the Design Center. Furthermore, the meticulous and thorough intersection control interface can be achieved for alternative and coordinated specialties. For example, in the design control of the prototype room at the sales office, the

2. 全程管控、要点突出

万达公馆、华府设计管理为全专业全过程管理，从项目前期产品定位、方案设计、方案深化、施工图到设计封样、施工实施，设计中心各专业均有详细完备的管控办法，并且各专业穿插配合、交叉界面管控均能做到细致深入。例如，在售楼处样板间的设计管控中，设计中心不仅承担设计管控工作，同时也承担了整个卖场的实施品质管控及进度计划管控。2013年全年的售楼处样板间均按时开放，品质优良率达到90%以上（图2、图3）。

3. 管理标准、高效务实

采用总表（设计模块化节点工作依据与设计成果一览表）、各阶段综合评审表、设计要点审核表的三级表格，全面覆盖整个开发流程。设计各阶段的各专业、各业态均列出了相应设计要点，明确了万达集团公馆、华府的基本要求和强制性标准，是设计方、项目公司、设计中心统一的技术标准（图4）。

设计各阶段设计节点均需进行三级评审，填写相应的《评审表》、《设计要点审查表》，以OA上报并审批通过为管理动作结束的标志；评审分为初审、复审、审定三个分项动作，分别由项目公司规划副总、设计中心负责人及标准组、设计中心总经理签字确认完成（图5）。

Design Center takes in charge of the work of design and control together with implementation quality control and process management of the whole sales place. All prototype rooms in sales offices were open on time in the whole year of 2013 and the quality excellence rate is over 90% (Figure 2, Figure 3).

3. STANDARD, EFFICIENT AND PRACTICAL MANAGEMENT

There are tables cover the whole development process in three levels, including Summary Table (Table of Work Bases and Design Outcome for Design Modular Milestones), General Assessment Table at each stage and Approval Table of Design Essentials. The design essentials are listed correspondingly for specialty and industry at each design stage to clear the basic requirements and mandatory standards of Wanda Properties for Sale, which is the uniformed technical standard for designers, project companies and the Design Center (Figure 4).

Design milestones at each design stage are required for review in three levels by correspondingly Assessment Table and Approval Table of Design Essentials and submitting them via the OA system to be approved as the end of the management flow. The assessment is divided into three stages: preliminary review, assessment and approval, signed correspondingly by Vice-manager on Planning of the Project Company, Project Manager in the Design Center, and General Manager of the Standard

（图2）沈阳奥体万达公馆建筑

（图3）大连高新万达公馆建筑

（图4）销售物业设计管控计划流程一览表

（图5）设计要点审查表

4. 考核坚决、正向激励

一方面建立对项目公司的评价和考核体系，从设计评审、设计管控、样板段质量、入伙检查四个方面进行评价，并作为年终考核的依据；另一方面建立设计中心内部评价考核体系，主要从节点按时完成情况、各阶段设计质量及实施效果，对设计中心项目经理实行正向激励。

同时，对供方设计院图纸质量及配合度进行履约考核，确保供方资源提供良好设计服务。

5. 整合资源、树立标杆

万达每年新增项目10~20个，为确保项目品质及设计快速推进，在提高设计效率的同时避免品质风险，设计中心每年投入巨大精力进行产品研发，通过不断发掘并整合各地项目实操中积累的丰富经验与资源，对具有代表性的产品进行标准化分析总结，以此建立产品标准及推进产品创新。2013年设计中心制定的户型模块、立面模块、售楼处模块、公寓写字楼模块等产品标准当年形成成果，当年推广应用，在激烈的市场竞争中历经考验，为项目的高效快速开发、完成当年销售指标打下了良好的基础（图6~图8）。

6. 因地制宜、突破创新

万达项目遍及全国各地，各地市场情况、客群的喜好、生活习惯各不相同，在销售物业设计管控中，设

Team and the General Manager of the Design Center (Figure 5).

4. RIGOROUSLY EVALUATION AND POSITIVE INCENTIVE

On one hand, the evaluation and examination system is established for Project Company in four aspects including design review, design control, prototype section quality and admission examination, which are taken as examination standard at the end of the year. On the other hand, internal evaluation and examination system in the Design Center is established to implement positive incentive for project manager of the Design Center mainly from timely completion of milestones, design quality and implementation effect at each stage.

At the same time, performance appraisal of drawing quality and cooperation is conducted for the design firm to guarantee excellent design service from the supplier.

5. INTEGRATING RESOURCES AND ESTABLISHING BENCHMARKS

There are 10 to 20 new projects in Wanda Group each year. To guarantee project quality and design progress, and prevent quality risk while improving design efficiency, the Design Center makes great efforts for product research and development each year. Standardization analysis and summary for representative products are made to set up product standard and promote product innovation by continuously exploring and integrating accumulative practical experiences and resources of projects in various places. Product standards, including

（图7）长沙开福万达公馆中轴景观

（图6）大连高新万达公馆样板间大堂

(图8)武汉K4万达公馆精装样板间

计中心与营销部、项目公司紧密配合,认真深入分析、研究当地市场、目标客群及竞品项目产品特点,制定针对性的产品策略,做到了项目产品既有标准化的细致,又有各自本土化的营销卖点。

设计中心在公馆、华府设计管控的实际经验中结合万达企业特点,形成了以上管控方法,并以此在既往工作中取得了一定的成绩。未来,设计中心仍将不断优化管控办法、不断进行产品创新,以确保在激烈的市场竞争中持续打造富有竞争力的优质产品,为营销指标的完成打下坚实的基础,同时继续赢得市场口碑及美誉度。

house type module, façade module, sales office module, apartment and office building module, etc. have been successfully formulated and applied by the Design Center in 2013. Having experienced in the fierce market competition, the above modules laid good foundation for the projects' efficient and rapid development and sales targets completion in that year (Figure 6 to Figure 8).

6. ADAPT TO LOCAL CONDITIONS AND INNOVATION BREAKTHROUGH

Wanda projects are distributed all around China. For different market conditions, favors and living habits of clients, the Design Center, closely coordinating with Marketing Department and Project Companies, makes careful and deep analysis of the local market, target clients and characteristics of the competitive project products in design control of Properties for Sale to work out specific product strategies. Thus, projects enjoy both meticulous standard and local selling point.

The Design Center has formed above control methods in the practical experience of design management of Properties for Sale combined with characteristics of Wanda Group. Achievements have been made in the past works with above methods. In the future, the Design Center will continuously optimize control methods and create product innovation to launch competent products in fierce market competition, lay solid foundation to reach sales targets and win praise and reputation in the market.

DALIAN DONGGANG WANDA MANSION
大连东港万达公馆

入伙时间	2010 / 10
建设地点	辽宁 / 大连
占地面积	6.52公顷
建筑面积	48.0万平方米

ADMISSION TIME	OCTOBER / 2010
LOCATION	DALIAN / LIAONING PROVINCE
LAND AREA	6.52 HECTARES
FLOOR AREA	480,000 m²

ARCHITECTURAL PLANNING
建筑规划

大连东港万达公馆项目北面大海，紧邻规划中的邮轮、游艇码头及达沃斯会议中心，总占地面积为6.52公顷，总建筑面积约为48.0万平方米。以营造高效的生态环境为要旨，万达公馆错落布置了3栋超高层公寓，保证了观海效果最大化。其中万达公馆销售部分为顶级国际公寓，是3栋185米超高层精装修板式楼。户型设计方面充分考虑动静分区、主仆分离。

The mansion is facing the ocean in the north and close to planning dock for cruise ships and yachts and Davos Conference Centre with land area of 6.52 hectares in total and an overall floor area of 480,000 square meters. Creating a high efficient ecological environment is the main idea of this project. Therefore, Wanda Mansion project contains three super high-rise apartments in a staggered layout to maximize the sea view. The Properties for Sale of which are three top-class international apartments of 185 meters in height. They are all slab-type buildings with fine decoration and fully consideration of separating living and private areas in the housing design.

1 大连东港万达公馆小区中心水景
2 大连东港万达公馆小区入口大门

1

2

SOFT DECORATION
软装

以华丽的装饰、浓烈的色彩、精美的造型,达到雍容华贵的装饰效果。以沉醉奢华,营造出和谐温馨、华贵典雅的居室氛围。卧室强调表面装饰,多用带有图案的壁纸、地毯、窗帘、床罩及帐幔等装饰画或物件。为体现华丽的风格,家具、画框的线条部位饰以金线、金边。房间采用反射式灯光照明和局部灯光照明,置身其中,舒适、温馨的感觉袭人,让那些追求品质生活的人找到了归宿。

This project adopts gorgeous ornaments, strong color and exquisite modeling to create dignified and graceful decoration effect and harmonious, sweet and luxury living atmosphere. The bedroom design is concentrated on surface decoration which uses patterned wallpaper, carpet, drapery, bedspread, valance and other decorative paintings and ornaments. To present a splendid style, furniture and picture frames are purfled with gold threads and golden edges. The warm feelings are aroused when clients are covered in the lighting reflection and partial lighting to make clients who pursue life quality feel like staying at home.

4

5

6

7

3 大连东港万达公馆内装旋转楼梯
4 大连东港万达公馆内装餐厅
5 大连东港万达公馆内装客厅
6 大连东港万达公馆内装客厅
7 大连东港万达公馆户型图

FUJIAN QUANZHOU WANDA MANSION
福建泉州万达公馆

入伙时间	2013 / 05
建设地点	福建 / 泉州
占地面积	18.33 公顷
建筑面积	123 万平方米
ADMISSION TIME	MAY / 2013
LOCATION	QUANZHOU / FUJIAN PROVINCE
LAND AREA	18.33 HECTARES
FLOOR AREA	1,230,000 m²

PROJECT OVERVIEW
项目概述

泉州浦西项目的总规划用地面积18.33公顷，规划总建筑面积123万平方米，其中地上91万平方米，地下32平方米。住宅区60万平方米，分成南北两区，共8栋楼，均为47层超高层高级住宅。

The overall planning land area of Quanzhou Puxi Project is 18.33 hectares, and its floor area is 1,230,000 square meters including 910,000 square meters above ground and 320,000 square meters underground. The residential district, which covers 600,000 square meters, is divided into north and south parts with a total of eight high-rise and high-grade residential buildings of 47 floors.

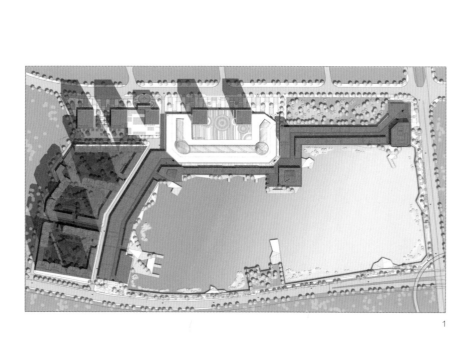

1

1 福建泉州万达公馆总平面图
2 福建泉州万达公馆建筑外立面
3 福建泉州万达公馆建筑顶部细节

5

ARCHITECTURAL PLANNING
建筑规划

项目规划采用U形布局，最大化利用江景、湖景；半围合大庭院提供舒适的公共空间和室外风景，既能充分享受到自然阳光、又可内院观赏湖景。Art Deco的建筑风格强调竖向线条，呈现出挺拔向上的气势。整体化的风格，使得高密度建筑群对城市肌理的侵扰被弱化，仿若滨江边的一块纯洁幕布，对泉州城市环境提升颇有贡献。

A U-shape layout is applied in this project to maximize the river and lake view. Half enclosure courtyard design provides pleasant public space and outdoor scenery for residents to make them fully enjoy natural sunlight and beautiful lake view. The Art Deco style emphasizes vertical lines and shows tall and straight imposing manner. The integrated style weakens the sense of intrusion towards urban texture caused by high-density building groups, and makes the mansion stand like a piece of pure curtain of the river. This project makes great contribution to promote the urban environment of Quanzhou City.

6

4 福建泉州万达公馆入口大门
5 福建泉州万达公馆顶部泛光
6 福建泉州万达公馆夜景

7

LANDSCAPE
景观

城市最后的奢侈莫过于自然，并非所有人都能坐拥自然美景。公馆秉承饱览人生的丰沃积淀和胸怀世界之广博境界，瞰环城水系，拥有360度观江、品湖之景观视野，占据城市资源与自然资源的优势。在万达公馆，可以享受到比肩世界顶级豪宅的"藏于世界中心，拥揽自然美景"的高端生活方式。

Nothing in urban is more luxury than nature. Not everyone could be in its favor. This mansion, with abundant accumulation of experiencing life and encyclopedic state of embracing the world, possesses 360 degree view angles of river and lake, and occupies advantages of urban resources and natural resources. Living in Wanda Mansion, residents could enjoy the upper-scale life style of "hiding in the center of the world, embracing the beauty of the nature", just like dwelling in a world-class deluxe villa.

7 福建泉州万达公馆核心景观
8 福建泉州万达公馆中心水景
9 福建泉州万达公馆园林

INTERIOR DESIGN
内装

本案以欧式古典风格元素贯穿整个项目，辅以石材、木材、壁纸、金箔等材料，体现整个豪宅项目的奢华品质。室内设计以对称手法突出了豪宅的尊崇感，尤其入口玄关的四根古典柱式更增加了仪式感，也成为该项目的一个突出卖点。为使方案做到主次有序，把设计重点放在了日常活动较多的玄关和客餐厅，墙面以石材和木作造型加以装饰。卧房区域只对主卧室进行了重点设计，床头的造型主墙以皮革和木作搭配，其余空间作简化处理。

European classical style, accompanied with stones, carpentry, wallpaper, gold foil and other materials, is applied in this project to incarnate the luxury character of the whole mansion. Interior design stresses the sense of veneration of the mansion by using symmetrical technique and especially four classic pillars at the hallway to enhance the sense of ceremony. This could be an appealing selling point for this project. To make clearly primary and secondary orders, the design concentrates on places with more daily activities, like hallway, living and dining room, and decorates the wall with stones and carpentries. For bedroom area, the design only aims at master bedroom with leather and carpentries on the main wall of the head of bed. Simplified designing is applied to other places.

10

11

10 福建泉州万达公馆大堂
11 福建泉州万达公馆客厅、过道、餐厅
12 福建泉州万达公馆客厅
13 福建泉州万达公馆户型图

12

13

CHANGSHA KAIFU WANDA MANSION
长沙开福万达公馆

入伙时间	2013 / 11
建设地点	湖南 / 长沙
占地面积	32784平方米
建筑面积	地上22.65万平方米
	地下5.9万平方米

ADMISSION TIME	NOVEMBER / 2013
LOCATION	CHANGSHA / HUNAN PROVINCE
LAND AREA	32,784 m²
FLOOR AREA	226,500 m² ABOVE GROUND
	59,000 m² UNDERGROUND

PROJECT OVERVIEW
项目概述

地块位于长沙市开福区五一大道与湘江大道交叉处以北，A区用地南北长200米，东西宽255米，用地面积3.28公顷，区内建42层、46层超高层高级住宅，地下设三层车库。

The project is located at the north of intersection of Wuyi Avenue and Xiangjing Avenue, Kaifu District, Changsha City. The land area of Zone A is 3.28 hectares, 200 meters long from north to south and 255 meters wide from east to west. There are high-rise and high-grade residential buildings of 42 floors and 46 floors with three levels of underground garage.

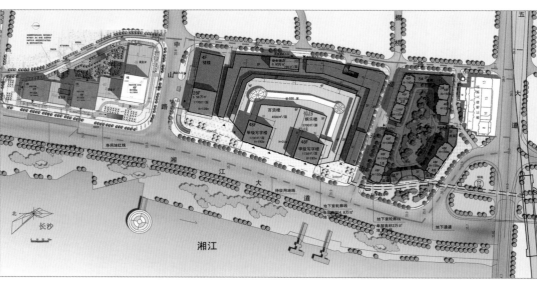

2

1

3

1 长沙开福万达公馆总平面图
2 长沙开福万达公馆中轴水景平面图
3 长沙开福万达公馆中轴景观夜景

ARCHITECTURAL PLANNING
建筑规划

公馆住宅建筑为竖向挺拔的Art Deco风格，采用纯净统一的石材及铝板，立面外观庄重俊朗。体量以双拼短板为主要形态，通过42~46层的高低错落，整体立面恢弘大气并充满活力，创造了沿江优美的天际线，并有效减小了建筑对城市江面的压力；建筑体量与开口虚实恰当，保证了日照均匀充分；公建化的沿街形象与B区商业呼应，造就挺拔优雅的城市综合体形象。

The mansion adopts vertically towering Art Deco style with pure and unified stones and aluminum sheets to make the facade appearance solemn and grand. Its volume mainly uses short plates with double splice. Through height difference between buildings from 42 floors to 46 floors, the whole facade looks majestic and magnificent and is full of dynamic, creating a polished skyline along the river and effectively decreasing the stress of urban river surface caused by buildings. The building volume is appropriate with opening, which guarantees evenly distributed and sufficient sunlight. The public buildings street image is in concert with section B commercial area to build up a stiff and elegant urban complex image.

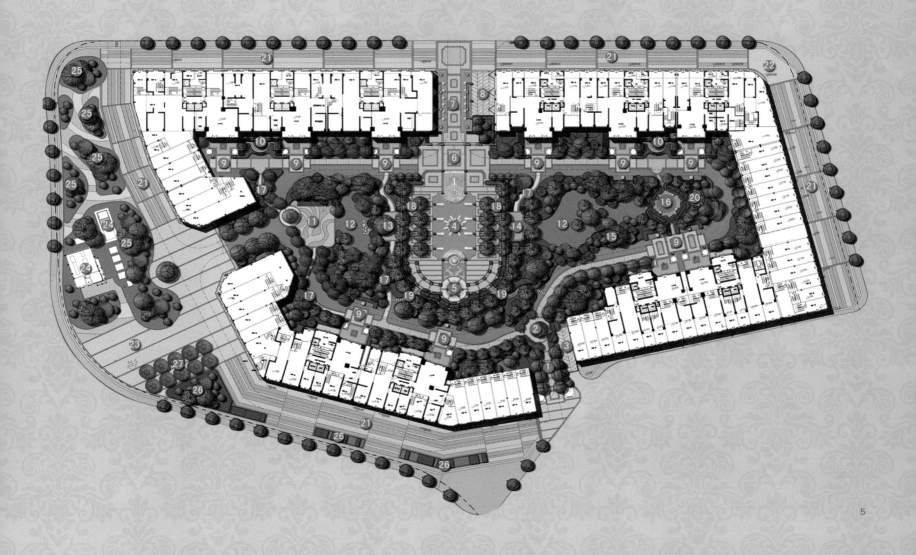

LANDSCAPE
景观

整体设计通过对场地性质、空间及周边环境的分析与总结，引入了尊贵、典雅、浪漫的法式大喷泉，结合强烈有序、层层渐进的轴线景观，强调秩序感和纵深感，体现出庄严的仪式感。景观融注了中世纪欧陆风格的人文关怀，体现了欧式经典的纯美。景观、建筑与整体环境风格和谐统一，同时满足了高层建筑的俯瞰效果及置身其中的体验感受。铺装材料、设计语言符号与建筑立面保持统一。

Through research and analysis of site, space and surrounding environment, the overall design brings in French-style great fountains of dignity, elegance and romance, integrating with axis landscape of intensive orders and progressive layers to present senses of sequence, verticality and solemn ceremony. The landscape reflects chaste and beautiful European classic style by melting humanistic care of euro-continental style from the medieval time. The landscape, buildings and entire environment are harmonic and unitized in a whole, and at the same time, meeting the demands of bird's-eye view from high-rise building and experience of involvement. Paving materials, design language symbols and building facade maintain uniform along with each other.

4 长沙开福万达公馆中轴雕塑
5 长沙开福万达公馆景观总平面图

引风景入内，隔喧嚣于外，花园入口大门无论从朝向、材质，还是尺度、细节等均非常讲究——石材的材质、颜色与建筑相融合，造就磅礴气势；大门的最外跨借用了两边底商的空间，在视觉上强化了入口的体量感；欧式铁艺大门的精致细节将考究的工艺展现得淋漓尽致；配以点缀式的暖色灯光，夜间的入口更别有一番景致。

Bringing in scenery and staying away from noise, the entrance of the garden is quite exquisite from whatever orientation, texture, height or details. Texture and color of stones blend with the building to express majestic and imposing momentum. Sense of volume of the entrance is visually reinforced when the gate's outmost span borrows space from floor traders of both sides. Delicate details of European-style iron gate perfectly demonstrate its fine process. Embellishing with warm color lighting, the entrance looks more distinct and distinguished at night.

8

9

10

6 长沙开福万达公馆主入口大门
7 长沙开福万达公馆主入口大门效果图
8 长沙开福万达公馆主入口铁艺门徽章设计
9 长沙开福万达公馆主入口铁艺门徽章效果图
10 长沙开福万达公馆主入口铁艺门构件设计
11 长沙开福万达公馆主入口铁艺门线稿图

11

次入口也是展示小区的一个窗口。其与正门的连线形成了小区的主要景观轴线。石材颜色与建筑颜色相融合，与主入口采用同样的材料，铁艺大门同主入口大门协调，制作精美、气势宏伟。

Also, the secondary entrance is a window of showing the community. It lines up with the main entrance to build axis landscape of the community. Stone color is coordinated with building color, adopting the same materials with the main entrance which is harmonious with exquisite and magnificent iron gate.

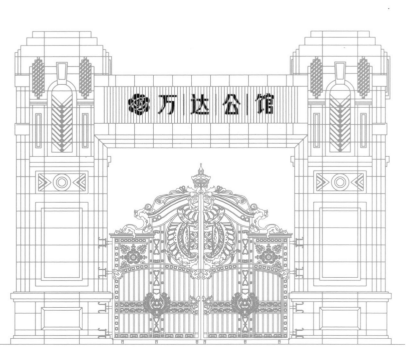

16

景观设计着重豪宅的私密性和园林的参与性。种植风格自然轻松，采用大量绿植。上层通过大面积树冠遮蔽的手法，打造隐秘的豪宅氛围；下层减少绿化层次，主要用地面植被的手法提高视线的通达性和亮度，以适应居住者活动和交流的需求，也营造出公馆大气宁静的空间感。

The landscape design concentrates on privacy of the mansion and involvement of the garden with an easy and natural planting style by adopting plenty of green plants. The upper part is to create covert atmosphere of the mansion with tree crowns in a large area, while the lower part is to reduce greening layers and to raise accessibility and brightness of view by planting ground vegetation to meet the demands of activity and communication and also generate sense of serene spaciousness of the mansion.

12　长沙开福万达公馆主入口铁艺门构件设计
13　长沙开福万达公馆次入口大门
14　长沙开福万达公馆主入口铁艺大门效果图
15　长沙开福万达公馆主入口立面图
16　长沙开福万达公馆活动场地景观绿化组团

17

18

19

夜景通过对冷暖光源的研究，精确地控制照度和色彩关系：水下灯采用冷光源（色温4500K），庭院灯、草坪灯等采用暖光源（色温2800K），其他采用偏白的暖光源（色温3000K），使景致呈现出晶莹、梦幻的氛围，形成情景交融的效果。

For nightscape, the relationship of illumination and color is precisely controlled by the study of cold and warm light sources. Underwater lighting adopts cold light source with color temperature of 4500K. Warm light source of 2800K color temperature is applied to garden lamps, lawn lamps and so forth. Other places adopt near-white warm light source of 3000K color temperature to create glittering and dreamlike atmosphere for the landscape and reach perfect harmony feelings.

17　长沙开福万达公馆中轴水景
18　长沙开福万达公馆单元入户景观
19　长沙开福万达公馆入口夜景
20　长沙开福万达公馆单元入户景观
21　长沙开福万达公馆花钵

充分考虑行人在步行中的行为模式，道路两边起微坡。结合地形的高差变化，在种植上运用重复、对比等手法，强调层次的丰富性。注重灌木的流线性，营造出"密不透风、疏可跑马"的绿化空间环境。通过开放空间和高低错落、层次感丰富的私密植物空间的对比营造，再加上丰富的地形变化，行走其间便有了置身于森林峡谷般的感受。

Through comprehensive consideration of pedestrian behavior, both edges of road are designed with slight gradient higher than the middle part. Integrating with altitude differences of the ground, planting techniques are applied to emphasize repetition and comparison to match the layers and heights. The landscape emphasizes the sense of flow of shrub to create changing environment of the grove of "densely spaced in one place, while sparsely spaced in the other place". Thus, the feeling of being in forest and canyon is generated when one walks in the garden.

22

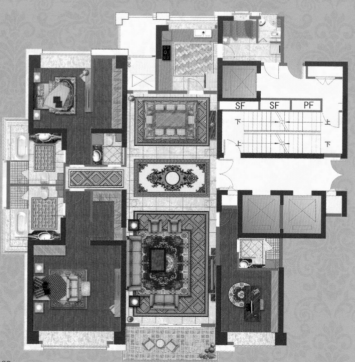

23

22　长沙开福万达公馆客厅
23　长沙开福万达公馆户型图
24　长沙开福万达公馆入户大堂
25　长沙开福万达公馆电梯间

24

INTERIOR DESIGN
内装

户内玄关部分设有四根罗马柱,既突出玄关部分,增强仪式感,又巧妙地把门厅和客厅分隔,玄关走道尽头设置精美的木雕。玄关地面采用大理石多色拼花;花型部分是紫罗红、金年华和西班牙米黄大理石,波打线是金年华和黑金花。吊顶采用椭圆雕花造型,配合金箔贴面,天地合一,显得既大气又奢华。客厅的沙发背景运用了经典柱式分隔,并附以装饰柱及雕花,天花造型融入洛可可风格的细腻精巧。

Four marble pillars in the hallway protrude the entrance part and reinforce the sense of etiquette, also skillfully separate the hallway from the living room. At the end of the hallway there are exquisite wood carvings. The hallway floor adopts marble multicolor parquet with image of Rosso Levanto flowers, Indus Gold and Spanish beige marble and boundary part of Indus Gold and Portopo. Blending in delicate rococo style, the ceiling is designed in oval carving patterns coated with gold foil to reveal magnificence and luxury and unify the sky and the earth. The sofa background in the living room is in European classic style with decorative columns and carving patterns.

25

WUHAN JIYUQIAO WANDA MANSION
武汉积玉桥万达公馆

入伙时间	2013 / 10
建设地点	湖北 / 武汉
占地面积	103200平方米
建筑面积	地上389400平方米
	地下87100平方米

ADMISSION TIME	OCTOBER / 2013
LOCATION	WUHAN / HUBEI PROVINCE
LAND AREA	103,200 m²
FLOOR AREA	389,400 m² ABOVE GROUND
	87,100 m² UNDERGROUND

ARCHITECTURAL PLANNING
建筑规划

项目位于武昌临江大道，充分利用临江的景观自然优势，为住户提供优质的观赏面及良好的日照条件。建筑设计风格为Art Deco，线条形式在竖向原则上灵活运用，通过重复、对称、渐变，凸显了立面的尊贵感。超高层高级住宅采用围合的结构，提供了优质安全的大景观庭院。

This project is beside Linjiang Avenue, Wuchang District. Taking full nature advantages of Linjing area, this project provides superior view and sufficient sunlight for residents. The architecture design adopts Art Deco style which concentrates on flexible application in vertical principle through techniques of repetition, symmetry and gradients to demonstrate the sense of solemnity of the facade. Enclosing structure is applied to the high-rise and high-grade buildings to provide superior and safe courtyard with macro-landscape.

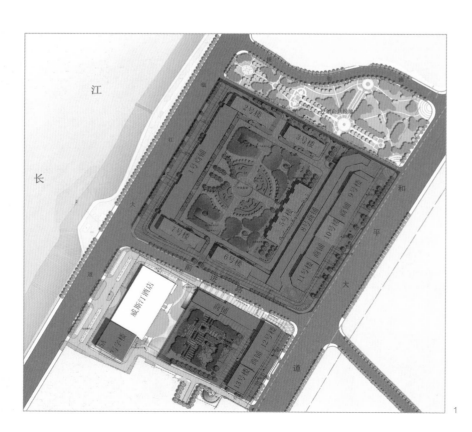

1

2

1 武汉积玉桥万达公馆总平面图
2 武汉积玉桥万达公馆建筑外立面
3 武汉积玉桥万达公馆主入口景观

4 武汉积玉桥万达公馆中轴水景
5 武汉积玉桥万达公馆主题雕塑
6 武汉积玉桥万达公馆景观亭廊

LANDSCAPE
景观

强调以华丽的装饰、浓烈的色彩、精美的造型，达到雍容华贵的新古典风格。在视线上，注重强化轴线的仪式感；景观小品上，注重衬托建筑的庄重感；景观空间上，注重产生引导感；植物营造上，注重不同功能分区的序列感及丰富植物的层次感，使建筑语言与景观元素互相融合。铺装材料、设计语言符号与建筑立面保持统一。

The landscape is designed in neoclassical style with gorgeous decoration, strong colors and exquisite modeling. For visual effect, it puts emphasis on the sense of ceremony of axis; for featured landscape, it concentrates on the sense of solemnity of building foil; for landscape space, it focuses on generating the sense of guidance; for plant arrangement, it pays close attention to the sense of sequence of different functional areas and the sense of layer of plentiful plants to mix in architectural language and landscape elements. Paving material, design language symbols and building facade maintain uniform with each other.

INTERIOR DESIGN
内装

设计表现上更多地体现了皇室般的艺术需要，追求唯美主义的真实体现；内装造型细腻，空间分割精巧，层次丰富，是装饰美与自然美的完美结合。简洁、明晰的线条勾勒出欧洲新古典的神韵，让人感受到恬静典雅，悠闲舒适。白色调与深啡色调的强对比，营造一种高贵雅致的空间深度感，将欧洲新古典风格演绎成一种优雅的平衡姿态，演绎出含蓄、考究的家居空间。

The interior design represents art needs of royal style and pursues realistic embodiment of aestheticism. It is a perfect combination of decorative beauty and natural beauty by adopting fine modeling, exquisite space partition and rich layers. Plain and distinct lines sketch romantic charm of European neoclassical style to make people feel peaceful, leisurely and comfortable. The sharp contrast of white and dark brown creates a sense of space depth of dignity and grace and presents exquisite living space by interpreting the European neoclassical style in a more humble way.

7 武汉积玉桥万达公馆精装样板间客厅
8 武汉积玉桥万达公馆精装样板间主卧
9 武汉积玉桥万达公馆入户大堂

JINAN WEIJIAZHUANG WANDA MANSION
济南魏家庄万达公馆

入伙时间	2010 / 09
建设地点	山东 / 济南
占地面积	23.0 公顷
建筑面积	100 万平方米

ADMISSION TIME SEPTEMBER / 2010
LOCATION JINAN / SHANDONG PROVINCE
LAND AREA 23.0 HECTARES
FLOOR AREA 1,000,000 m²

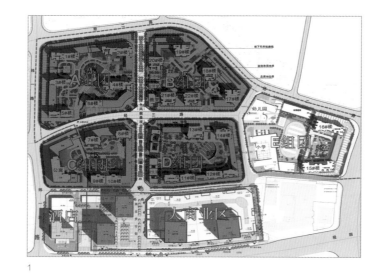

1

PROJECT OVERVIEW
项目概述

济南万达广场魏家庄项目占地面积23.0公顷，总建筑面积达100万平方米。其中大商业面积16.0万平方米，室内步行建筑面积达4.0万平方米，提供近3000个停车位。项目包括顶级购物中心、文化娱乐休闲设施、公共空间、5A级写字楼、超五星级酒店和高级公寓等，集购物、休闲、餐饮、文化、娱乐、会展、商务、停车、酒店、办公、居住和旅游等多种功能于一体，从而形成城市商业中心、生活中心和财富中心。

The project land area is 23.0 hectares and its overall floor area is 1,000,000 square meters, including large commercial area of 160,000 square meters and indoor walking floor area of 40,000 square meters. This project contains top-class shopping mall, entertainment, recreation facilities, public space, 5A office building, superior five-star hotel, premier apartments, nearly 3,000 parking lots, etc.. Integrated with the functions of shopping, relaxation, dining, entertainment, exhibition, business, parking, hotel, office, residence and tourism, this project is forming a new urban commercial center, life center and fortune center.

1 济南魏家庄万达公馆总平面图
2 济南魏家庄万达公馆商业街效果图
3 济南魏家庄万达公馆景观平面图
4 济南魏家庄万达公馆室外亭

2

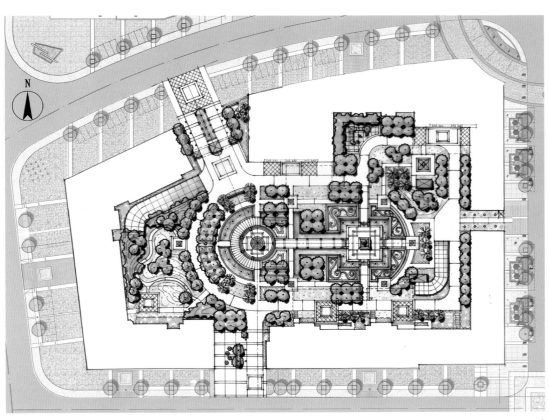

ARCHITECTURAL PLANNING
建筑规划

建筑以体块简洁、色彩明快、注重细部处理和材质对比的手法，取得了鲜明的个性。项目的立面处理，采用简约现代风格，结合小区氛围运用不同的材质组合与色彩对比，形成既自然又不乏条理的型体特征；立面造型通过玻璃与阳台的组合，简洁大气中透露丰富的细节。

The bright and distinct character of the architecture is benefited from clear lines, lively color, concentration on detail treatment and texture comparison. The facade is designed in simplified modern style with community atmosphere by adopting various material combinations and color comparison to display natural and organized features. Facade modeling shows rich details in the plain and imposing design through combination of glass and balconies.

LANDSCAPE
景观

将欧洲传统文化特色与中国地方文化相结合，运用富有生命力的文化底蕴去营建景观。设计线条曲直结合，空间开合得体，充分贯彻"以人为本"的思想，提供不同性质、功能、尺度的交往空间。以景观生态学的理论为指导，设计优化人工植物群落，运用植物的多样性及各种形态、形式，有效地增加了绿量，提高了生态效益，改善了环境质量，充分发挥绿地对居住环境的改善作用。

Combining European traditional culture with Chinese local culture, the landscape is designed from lively and rich cultural deposits. With appropriate and suitable lines and space, the design intends to comprehensively implement the idea of "human oriented" and provide communicative space of different characters, functions and sizes. Guided by landscape ecology theories, designer optimizes the layout of plant community by utilizing diversity of plants and its various vegetative stages to effectively enlarge green space, remarkably increase ecological benefits and greatly improve environmental quality for the living environment improvements.

INTERIOR DESIGN
内装

首层入户大堂，以极简线条打造宫廷级居住空间。超长面宽的墙面造型采用米黄石材干挂，装饰画点缀；地面采用米黄、浅啡网与深啡网石材拼花。标准层一级吊顶标高2800毫米，二级吊顶标高2600毫米。室内装修面层材料，采用石材、地板、木线条、软包、壁纸、马赛克等。室内空间布局独到，南北通透，结构紧凑，功能完备；在照顾功能的同时，更注重朗阔感与舒适性。

Quite simplified stripe pattern is applied to the hallway of the room on the ground floor to create palace-like dwelling space. Wall surface modeling of over length face width adopts dry-hung beige stones with the embellishment of decorative paintings, and ground adopts stone parquets of beige color, light emperador and dark emperador. In typical floor, the elevation of the first suspended ceiling is 2,800 millimeters, and the second suspended ceiling is 2,600 millimeters. The surface materials of interior decoration include stone, floor board, wood line, soft background wall, wallpaper, and mosaic tiles, etc. The interior space layout is original and unique with good ventilation, compact structure and full functions. With satisfying functional requirements, the interior design lays more emphasis on senses of spaciousness and comfort.

5 济南魏家庄万达公馆入户大堂
6 济南魏家庄万达公馆户型图

7 济南魏家庄万达公馆餐厅
8 济南魏家庄万达公馆客厅
9 济南魏家庄万达公馆主卧
10 济南魏家庄万达公馆主卧

SHIJIAZHUANG YUHUA WANDA MANSION
石家庄裕华万达公馆

入伙时间	2013 / 04
建设地点	河北 / 石家庄
占地面积	40.5公顷
建筑面积	183.9万平方米
ADMISSION TIME	APRIL / 2013
LOCATION	SHIJIAZHUANG / HEBEI PROVINCE
LAND AREA	40.5 HECTARES
FLOOR AREA	1,839,000 m²

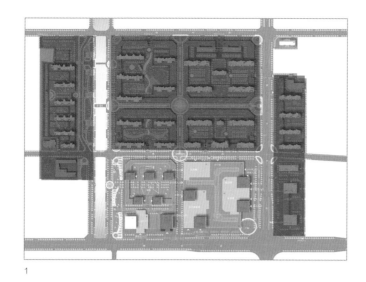

1

PROJECT OVERVIEW
项目概述

石家庄裕华万达广场位于石家庄市裕华区。地块范围东临建华大街两侧，西邻民心河两侧，南侧为槐安东路、世纪公园，北侧为槐中路。本工程规划用地面积约为40.5公顷，其中商业办公用地面积约为14.42公顷，总建筑面积约为183.9万平方米。

This project is located at Yuhua District, Shijiazhuang City, with Jianhua Street in the east, Minxin River in the west, Huaian East Road and Century Park in the south and Huaizhong Road in the north. The planning land area is approximate 40.5 hectares including commercial office area of around 14.42 hectares and overall floor area of about 1,839,000 square meters.

ARCHITECTURAL PLANNING
建筑规划

立面风格选择Art Deco风格，基于线条形式在竖向原则上灵活运用，通过重复、对称、渐变等处理，凸显立面的高大，给人以华贵高尚的尊贵感。立面材质以浅色为主，多采用米黄、咖啡色的天然石材，与玻璃幕墙的结合，形成虚实对比。整个造型经典、含蓄、富有内涵，是对当地文化精髓的提取和继承，也是创新开拓精神的展现。

The architectural facade adopts Art Deco style which bases on flexible application in vertical principle through techniques of repetition, symmetry and gradients to highlight grandeur of the facade and create senses of luxury and dignity for residents. The facade texture is mainly in light color such as natural stones in beige and coffee colors, integrating with glass curtain wall to form a virtual-real comparison. The entire modeling of classic, modesty and connotation is extraction and inheritance of quintessence of the local culture, and also embodiment of innovative and pioneering spirit.

2

1　石家庄裕华万达公馆总平面图
2　石家庄裕华万达公馆建筑外立面
3　石家庄裕华万达公馆景观总平面图
4　石家庄裕华万达公馆小水景墙

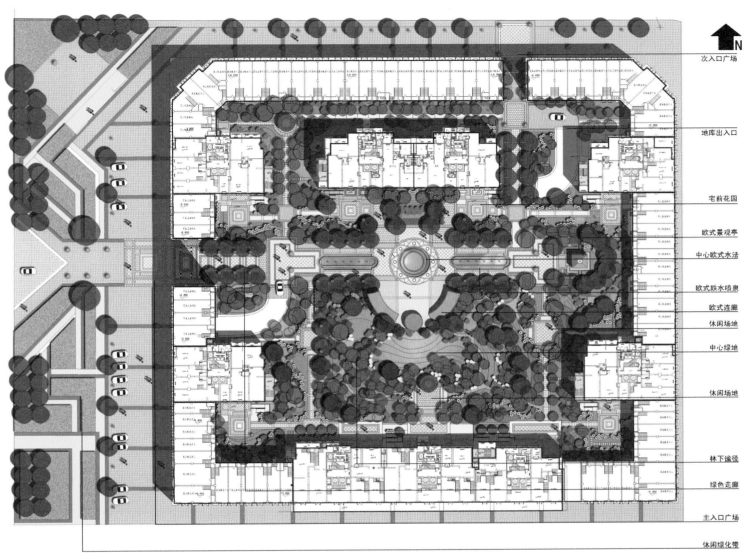

	次入口广场
	地库出入口
	宅前花园
	欧式景观亭
	中心欧式水法
	欧式跌水喷泉
	欧式连廊
	休闲场地
	中心绿地
	休闲场地
	林下蹊径
	绿色走廊
	主入口广场
	休闲绿化带

LANDSCAPE
景观

景观设计营造了多空间环境及不同的空间体验，尤其是园区的核心景观——中心水景区——用静水面、跌水、喷水水法等水景元素，构成了动感强烈的欧式水景广场。两侧的静水面与中心水法相呼应，形成磅礴大气而又不失品位的聚集空间。同时，此空间还具有整体园区重要的活动场所的功能。细部设计精雕细刻，运用多种石材与金属，刻画出尊贵、舒适的环境景观。

The landscape design creates multi-space environment and different space experience, especially the core landscape in the garden – center waterscape area – which adopts water features, such as still water, cascade and artificial fountain, to establish dynamic waterscape square of European style. The still water surfaces on both sides echo with the central fountain to form aggregate space of majesty and vast without losing the taste. Meanwhile, this space contains important functions for activities in the whole garden. The detail design is exquisite by applying various stones and metals to depict honorable and comfortable environmental landscape.

| PART B | WANDA MANSIONS & WANDA PALACES
万达公馆、华府

INTERIOR DESIGN
内装

金黄色的色调带来一种五星级酒店的奢华与尊贵，地面与墙面均采用大理石拼花砌成，墙面上金色腰线，增加层次感，吊灯为进口水晶灯；同时每个入户大堂都会设有管家式服务台，为业主24小时提供便捷的服务，私密而有安全保障。大堂的入户门，高约6米，宽4.8米，采用铜制。此门在设计风格上古典与时尚相结合，在色调上讲求沉稳、庄重与大气，能体现出尊贵与荣耀。

Golden color brings a tone of luxury and dignity of five-star hotel. Both floor and wall adopt marble medallion masonry with golden string courses on the wall and imported crystal lamp to increase the sense of layers. Moreover, steward type service desk is designed in every hallway of the apartment to offer convenient service in 24 hours for residents and make them feel private and secured. With 6 meters high and 4.8 meters wide, the brass-made main door in the hallway pursues a combination of classic and fashionable design style and stress calm, solemn and spectacularity in color to reflect its dignity and glory.

5

6

5　石家庄裕华万达公馆户型图
6　石家庄裕华万达公馆电梯间
7　石家庄裕华万达公馆入户大堂

9　　　　　　　　　　　　　　　　10　　　　　　　　　　　　　　　　11

餐厅（开间约4米，面积约15平方米）以椭圆形餐桌摆放，突出空间的和谐感，也契合一家人团团圆圆的温馨场面。南向双主卧的设计采光充足，更让家里暖意融融；卧室配有约8平方米的衣帽间，功能合理。地面均采用耐用的深胡桃色实木复合地板（地暖专用）。墙面铺壁纸，床头部位做布面硬包装，为房间增添了温馨的感觉，使整个房间充满生气。窗户采用隔音保温效果优异的中空玻璃，型材选用先进的断桥铝合金。

The dining room, with a bay of about 4 meters wide and floor area of about 15 square meters, sets an oval dining table to enhance harmonious sense of space which is also corresponding with sweet and warm scene of uniformity and happiness of a whole family. Two master bedrooms in the south direction provide sufficient sunlight to make home warm and comfortable, and each of them contains a coat room of 8 square meters with well-designed functions. The floor uses durable solid wood floor in dark nut-brown exclusively for floor heating. The wall and the head of bed respectively adopt wallpaper and hard decoration with cloth cover to increase senses of warmness and sweetness for the room and make the entire room lovely and lively. Windows are designed in advanced bridge-cut aluminum alloy and insulating glass with excellent sound insulation and heat preservation.

8　石家庄裕华万达公馆入户玄关
9　石家庄裕华万达公馆餐厅
10　石家庄裕华万达公馆主卧
11　石家庄裕华万达公馆主卧
12　石家庄裕华万达公馆客厅

12

ZHANGZHOU BIHU WANDA PALACE
漳州碧湖万达华府

入伙时间	2012 / 12
建设地点	福建 / 漳州
占地面积	14.39 公顷
建筑面积	59.8 万平方米

ADMISSION TIME	DECEMBER / 2012
LOCATION	ZHANGZHOU / FUJIAN PROVINCE
LAND AREA	14.39 HECTARES
FLOOR AREA	598,000 m²

PROJECT OVERVIEW
项目概述

项目定位为漳州市中高档休闲商住区，以商业、休闲为龙头，倡导和谐宜居的便利生活方式。住宅平行道路布置，庭院景观空间视线通透。建筑立面形象鲜明，具有强烈的视觉效果和标志性，将漳州碧湖万达广场打造成该区域的地标。

This project is positioned as a commercial and residential area of middle and top grades leading by business and relaxation industries to advocate harmonious and livable life style. Residential buildings are parallel to roads, and the garden view is full and clear. The building facade is distinctive, possessing excellent visual effect and symbolic role and making the Zhangzhou Bihu Wanda Palace a landmark in this region.

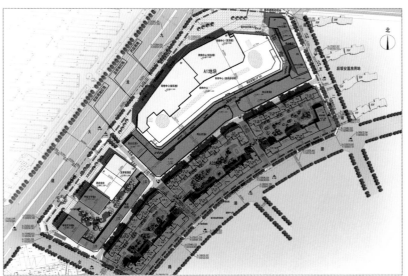

1　漳州碧湖万达华府外立面
2　漳州碧湖万达华府总平面图

ARCHITECTURAL PLANNING
建筑规划

漳州碧湖万达华府以简洁、高雅、挺拔的艺术线条呈现出Art Deco独一无二的建筑气质。项目在追求整体品质感的同时，也通过细部的营造，不动声色地宣扬着一种空间的美学，一种享受的境界。小区内采用人车分流设计，使得小区的安全、人居环境得到保障。主要的人流出口尽量避免设置在商业人流集中点和机动车来往线路。

Zhangzhou Bihu Wanda Palace adopts simple, elegant and towering line art to present unparalleled architectural temperament of Art Deco style. This project pursues the sense of quality, and in the meanwhile quietly propagates a kind of spatial aesthetics and a kind of enjoyable state through detail creation. The separation design of pedestrians and vehicles guarantees the community's safety and living environment quality. Main exits of pedestrian stream avoid crowds in business area and the route of vehicles.

4

LANDSCAPE
景观

场地建筑的格局为东西方向，空间呈南北向的狭长空间，小尺度、大景深的景观格局。通过竖向的变化营造出景观的特色，注重刻画地形变化及构建竖向景观元素（如跌水和构筑物）。

The architectural pattern is mainly on the east and west sides, and the space appears long and narrow from the north to the south to form a landscape pattern of small scale and large depth. Landscape design adopts vertical changes to create landscape features, focusing on depicting topographical changes and establishing vertical landscape elements, for instance cascades and structures.

5

6

3 漳州碧湖万达华府外立面
4 漳州碧湖万达华府建筑立面图
5 漳州碧湖万达华府水景
6 漳州碧湖万达华府景观庭院

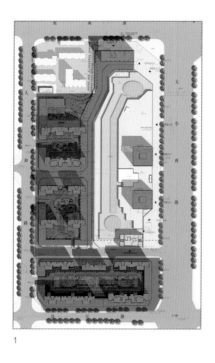

1

ZHENGZHOU ERQI WANDA PALACE
郑州二七万达华府

入伙时间	2013 / 07
建设地点	河南 / 郑州
占地面积	109962.27 平方米
建筑面积	43.86 万平方米

ADMISSION TIME	JULY / 2013
LOCATION	ZHENGZHOU / HENAN PROVINCE
LAND AREA	109,962.27 m²
FLOOR AREA	438,600 m²

1　郑州二七万达华府总平面图
2　郑州二七万达华府商业鸟瞰图
3　郑州二七万达华府商业街立面图
4　郑州二七万达华府住宅建筑外立面

ARCHITECTURAL PLANNING
建筑规划

住宅立面形象鲜明，采用经典简欧的设计手法，充分体现万达集团具有时代性的设计理念及产品的品质要求。建筑的头部做了一些简化的欧式构架造型，很好地活跃了高层建筑的天际轮廓线。建筑的基座采用了深色石材，墙身中段采用米黄色的仿石漆，整栋建筑显得尊贵大气。

Adopting classic simplified European style, the bright and vivid building facade fully displays design philosophies of modernity of Wanda Group and required high qualities of its products. Simplified European framework modeling is applied on the top of high-rise buildings, sketching out a brisk skyline. The whole building appears dignity and magnificent through dark stone foundation and beige stone-like paint at the middle of wall.

5

6

5 郑州二七华府中心庭院
6 郑州二七华府连廊
7 郑州二七华府小品花池
8 郑州二七华府景观
9 郑州二七华府绿化

LANDSCAPE
景观

采用法式风格，主轴线贯穿三个地块，根据地块体量形成不同的景观特色——A地块以人物雕塑为主；B地块以庭院式植物造景为主；C地块以次轴线对称种植为主。景观铺装规整、简洁大气，通过景观花池的宽大座椅配以地下防腐木地面，在商业广场的局部营造出园林的感觉；斜坡式花池与市政行道树间，形成幽长、别致的绿茵长廊，同时又将道路上行人的视线引导至商业建筑区的立面。

The landscape is designed in French style. The main axis penetrates the three blocks and form different landscape features according to block volumes: Block A focuses on figure sculptures; Block B focuses on garden-type plant landscape; and Block C focuses on symmetrical planting on the secondary axis. In commercial plaza, the orderly landscape pavement in coordinating with wide benches beside flora pools and antiseptic wood flooring create a feeling of walking in the garden. The space between gradient flora pools and street trees forms a long green corridor which leads pedestrians' sight to the commercial building facades.

FUJIAN PUTIAN WANDA PALACE
福建莆田万达华府

入伙时间	2013 / 09
建设地点	福建 / 莆田
占地面积	93086.95 平方米
建筑面积	31.39 万平方米

ADMISSION TIME	SEPTEMBER / 2013
LOCATION	PUTIAN / FUJIAN PROVINCE
LAND AREA	93,086.95 m²
FLOOR AREA	313,900 m²

PROJECT OVERVIEW
项目概述

住宅地块分为东西两块，由6栋32层住宅和1栋30层住宅组成，其中一二层为裙房商铺。

The residential area is divided into two sections, the east and the west, and consists of six buildings of 32 floors and one building of 30 floors. The ground and second floors of the buildings are podium stores.

ARCHITECTURAL PLANNING
建筑规划

建筑立面采用华贵、高雅的Art Deco风格，讲究庄重、华贵；同时摒弃过于复杂的肌理和装饰，迎合现代人的审美情趣。设计手法严谨、细节丰富、尺度宜人，创造出典雅高贵的居住建筑形象。

The building facade is designed in luxurious and elegant Art Deco style, focusing on solemn, luxury and exaggerated embellishment. In the meanwhile, it abandons complicated texture and decoration to cater new aesthetic taste of modern people. Rigorous design method, rich details and appropriate sizes create an elegant and noble image of the residential buildings.

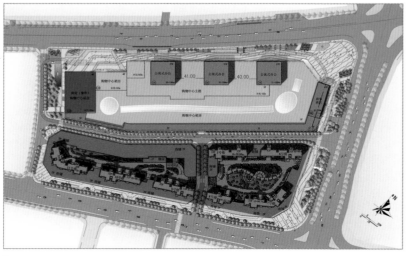

1 福建莆田万达华府总平面图
2 福建莆田万达华府建筑外立面

3

LANDSCAPE
景观

景观设计遵循以绿为锦，以水为绣，采用西式传统造园手法契合建筑主体风格，体现现代高尚住宅浪漫、高雅的格调。透过收放有致的雕塑水景空间，营造多变且富有韵律感的景观环境，辅以景观灯柱、多层次的绿化空间，为人们提供尺度亲切、生态和谐的高层次精神愉悦空间。

The landscape design concentrates on embellishment of green plants and water, using western traditional landscaping method to correspond to the main building style and establish romantic and elegant style for modern residential buildings. Well-arranged sculptures in waterscape space create varied and rhythmical landscape environment, aligning with lamp stands and multilayer green space as background to provide high-level pleasant spirit space of appropriate sizes and ecological harmony for residents.

3 福建莆田万达华府绿化
4 福建莆田万达华府喷泉
5 福建莆田万达华府水景夜景

PART C WANDA COMMERCIAL PLANNING
WANDA SALES PLACES
万达销售卖场

DESIGN TRILOGY OF WANDA SALES PLACES
万达销售卖场设计三部曲

文／万达商业地产设计中心技术管理部副总经理　付东滨

万达销售卖场作为万达最重要的销售工具，包含售楼处、样板间两个部分。在万达，销售卖场开放、商业开业、物业入伙并称三个核心事项。万达公馆销售卖场通常采用为欧式古典风格，万达华府销售卖场采用简欧、Art Deco、现代风格，通过制定完整的参观动线，对未来住宅环境空间的浓缩和提炼，再现高品质生活方式。

万达销售卖场工作围绕功能表现、价值挖掘和氛围营造三部曲展开。

第一部曲：功能表现

万达卖场强调视觉冲击功能的同时强调使用功能。其功能表现形式通过建筑、精装、景观、机电、结构等专业体现，万达卖场除了定义必要功能外，更加强调目标客户群可感受到的人性化功能。在售楼处、样板间的设计中也融入了特有的功能，如迷你高尔夫、室外水吧、室内空气净化系统、室内一键式照明开闭功能等（图1～图3）。

As the most important sales tool, Sales Places of Wanda are composed of sales offices and prototype rooms. Wanda considers the openings of sales places, the openings of commercial properties, and the property admissions as three core issues. Normally, Sales Places of Wanda Manson are in classical European style while Wanda Palaces are in simplified European style, Art Deco or modern style. By establishing visitors' movement patterns, it concentrates and extracts living space of the future and therefore represents high quality living style.

Wanda Sales Places carry out its work through three stages: function presentation, value exploration and atmosphere creation.

Episode I: Function presentation

Sales Places of Wanda emphasize on both visual impact and functions. Its functional presentation is reflected in the aspects of architecture, decoration, landscape, M&E, and structure. Besides the defined necessary functions, Sales Places of Wanda pay more attention to the humanized functions felt by the target clients. Specific functions are integrated into the design of sales offices and prototype rooms, including mini golf, outdoor wet bar, indoor air cleaning system and indoor one-button lighting switch, etc (Figure 1 to Figure 3).

I. ARCHITECTURE

To form attractive and powerful external visual impact, the buildings adopts door and canopy with great proportion, various façade styles and the according night lighting (Figure 4).

（图1）长沙开福万达公馆售楼处迷你高尔夫场

（图2）沈阳奥体万达公馆售楼处室外水吧

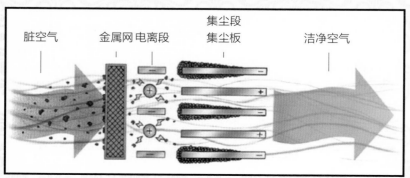

（图3）空气净化系统

（图4）沈阳奥体万达公馆售楼处

一、建筑

比例协调、气势磅礴的门头、雨篷，丰富的立面风格、考究的建筑比例、延续建筑风格的夜景照明，形成外在的视觉冲击（图4）。

二、室内精装

合理比例的开间、进深、净高打造出适宜的室内空间感，与属地文化结合的造型形成室内的视觉冲击。样板间设计中充分展示家居储物功能、场景灯光功能。普通小户型麻雀虽小，五脏俱全，设计中注重小空间下百变功能，一张床收起来是书桌，放下来兼具储物功能；小空间厨房整合抽拉式餐桌功能（图5）。

三、景观

销售卖场景观设计，首先要满足营销功能，提供开展营销活动所需要场地。通过对营销场地的景观环境设计，营造优美生态环境、柔化建筑立面空间。在提升项目的品质的同时，激发客户潜在购买欲望。

II. INTERIOR DECORATION

Feasible indoor space is presented in reasonable proportion of bay, depth and clear height. The shape combined with the local culture forms indoor visual impact. Storage function of the furniture and lighting function of the scene are fully displayed in the design of the prototype room. The general small house type still has complete functions in such a tiny size. The design highlights changeable functions in a small space that a bed is folded to become a desk and unfolded to store objects; what is more, there is a small space kitchen which integrates the function of withdrawing dining table (Figure 5).

III. LANDSCAPE

Landscape design in sales places firstly meets marketing requirements by providing place for marketing activity. The design of landscape environment in the marketing site creates beautiful ecological environment and smoothes the façade space of the building. Thus it will inspire the potential purchase desire of clients whilst improving the qualities of projects.

1. ENTRANCE
The entrance is used to display project style, which can affect people's mood. Since entrance in European style emphasizes the sense of sequence and ceremony, copper carving should be designed at steps of the entrance or on the ground close to steps to present magnificent palace style (Figure 6).

2. VISIT CHANNEL SURROUNDING THE PROTOTYPE ROOM
The visit channels surrounding the prototype room are affective impulse areas. Beautiful landscape all the way leaves deep impression on clients, bringing them higher expectation for visiting the prototype room. That promotes marketing function of the whole sales place.

（图5）一字型厨房+抽拉式餐桌

1. 入口区

入口为情绪酝酿区，展现项目的风格，欧式入口需强调序列感和仪式感，可在入口台阶或近台阶地面设置铜地雕，彰显华贵的宫廷风格（图6）。

（图6）烟台芝罘万达公馆售楼处前广场景观

2. 样板间周边参观通道

样板间周边参观通道：为情感冲动区，一路美好的景观让客户留下深刻印象，并对即将到来的样板间参观产生出较高的期待，使整个卖场的强化宣传功能得以实现。

3. 样板间窗景

样板间窗景：主要功能房间（主卧室，客厅，餐厅）窗景为重要设计点。推开窗的每一个空间都是景观重点，使得客户的感性冲动受到来自多方面的强烈冲击，一路积累，从而产生强烈的购买欲望（图7）。

4. 情景化小品

景观家具及小品设计，应与项目风格相同，注重装饰小品对空间气氛的烘托。

在功能表现中采用"两个基本工具"和"一项关键步骤"强化管控。"两个基本工具"是功能清单和流程图，用于对必要功能和人性化功能的梳理。"一项关

（图7）长沙开福万达公馆售楼处后花园景观

3. WINDOW SCENERY OF THE PROTOTYPE ROOM

Window scenery of the prototype room: window scenery is one of the most important design points for main functional rooms (master bedroom, living room and dining room). Every landscape should be able to bring impressive impact on clients and make them generate strong desire for purchase while opening the window (Figure 7).

4. SITUATIONAL ORNAMENTS

The design of landscape furniture and ornaments should be consistent with project style, emphasizing ornaments' decoration function in order to improve the atmosphere in the space.

"Two basic tools" and "one key step" are applied to enhance the control management in functional presentation. "Two basic tools" are functional list and flow diagram to arrange necessary functions and humanized functions. "One key step" is simulative use of design achievement, such as PPT to check and optimize accuracy of function positioning.

EPISODE II: VALUE EXPLORATION

I. EXCESSIVE COMPETITIVE MARKET

Having been intensively developed for almost twenty years, the real estate in China has finally entered into a mature stage. Competition is not only common but becomes increasingly fiercer. Wanda needs to think how to show its advantages in the competition and give clients a reason to purchase. The answer of how to make breakthrough is value exploration.

II. HIGHLIGHTS OF VALUE EXPLORATION

Taking interior decoration as an example, since the market capacity and purchasing ability in 3rd or 4th tier cites are not sufficient, the small house type with low total price is always a better choice in the market. Thus, how to equip small type with the feeling of a big room with complete functions becomes the most important design concerns of such type of houses. This kind of changeable space design is the result of value exploration. These two are mutually constrained and interacted (Figure 8).

键步骤"是指对设计成果模拟使用，如PPT，用于校核、优化功能定位的准确性。

第二部曲：价值挖掘

一、过度竞争的市场

中国的房地产经过了近二十年发展，已经进入成熟阶段，市场竞争成为常态化，并且日益激烈。如何在竞争中体现优势，给客户一个购买的理由成为万达的思考主题。如何突破，答案是价值挖掘。

二、价值挖掘的主要亮点

以内装为例，三、四线城市市场容量小，市场购买力不足，总价低的小户型必然是市场的一种选择。小户型如何让空间感觉大，功能齐备，容纳性好，要做到麻雀虽小，五脏俱全。小空间的百变空间设计就是价值挖掘的结果。两者互相制约、互相作用（图8）。

（图8）小户型百变空间：放下是床，收起是书桌

（图9）湘潭售楼处效果图

III. DESIGN PROCESS OF SALES PLACES OF WANDA IS A PROCESS OF VALUE EXPLORATION

Value exploration should be consistent with the market orientation and pay serious attention to cost control, construction difficulty and technology maturity at the meanwhile. Measures that generated from value exploration include excellence evaluation, summary, case sharing and experimental control. As the former two are the pathway towards value exploration, the latter two are promotion measures for value exploration achievements. Excellence evaluation is the process from quantitative change to qualitative change; summary is sublimation process of qualitative change and case sharing is the accumulative process from qualitative change to quantitative change again (Figure 9).

EPISODE III: ATMOSPHERE CREATION

I. CREATE ATMOSPHERE BY SATISFYING PSYCHOLOGICAL REQUIREMENTS OF CLIENTS

"Clay is molded into a vessel because of the hollow we may use the cup. Walls are built around a hearth because of the doors we may use the house. Thus tools come from what exists, but use from what does not", as quoted from Chapter XI of Lao Tsu: Tao Te Ching. The atmosphere of Sales Place of Wanda is created in five aspects which are the sense of sight, hearing, touch, smell and taste. Resonation is generated by shaping the scene in literature works (movie and fiction).

II. CREATE ATMOSPHERE BY DESIGNING

Many elements are adopted in order to satisfy clients' material, mental and psychological demands. These include the sense of space, tone, product function, publicity of themed culture, wonderful background music, comfortable temperature, moisture and clean environment, and advertising decoration and package. Thus creates a complete feeling environment for potential clients.

三、万达卖场的设计过程是一个价值挖掘的过程

价值挖掘坚持市场导向，兼顾成本、施工难易、技术成熟等方面。万达卖场价值挖掘形成了评优、总结、案例分享、试点管控等措施。评优、总结是价值挖掘的途径；案例分享、试点是价值挖掘成果的推广措施。评优是量变积累过程，总结是质变升华过程，案例分享是质变的再次量变积累过程（图9）。

第三部曲：氛围营造

一、通过满足客户心理要求营造氛围

老子《道德经》第十一章："埏埴以为器，当其无，有器之用，凿户牖以为室，当其无，有室之用。故有之以为利，无之以为用"。万达卖场的氛围营造从视觉、触觉、听觉、味觉、嗅觉五个方面着手。代入（场景）诸如文学作品（电影、小说），通过场景的塑造形成强烈的代入感，进而产生共鸣。

二、通过设计营造氛围

通过空间感、色调、产品功能、主题文化渲染、美妙的背景音乐，舒适的温度、湿度、洁净度，具有渲染力的美陈包装等众多元素，满足客户的物质需求、精神需求、心理需求，进而为潜在客户营造一个五味俱全的卖场环境。

1. 有基础（硬质）

材质、色彩、光（功能：物质需求）：在设计中，用质感效果突出的外立面石材、勾勒出建筑风格的夜景照明、拼花造型的石材地面、采光兼具保温的断桥铝合金窗体、仪式感强烈的水晶灯具及材质固有色彩属性的合理搭配，冲击客户的物质需求。

2. 有内涵（软质）

风格、主题（功能：精神需求）：万达卖场一直坚持文化属性、结合属地文化特点、主题与风格统一，形成冲击客户的精神需求。如大连高新万达公馆售楼处，通过对大连临海城市海洋文化的分析及定位，最终确定"海洋文化"主题。卖场开放时，背景音乐响起、名厨主理、美陈包装，共同营造卖场文化。

1. FOUNDATION (HARDNESS)
Texture, color and lighting (function: material demand): in the design process, we are trying to attract clients to purchase through the combination of various elements such as external façade stone with prominent texture effect, nightscape lighting in architectural style, parquet stone floor, AL window with lighting and insulation functions, ritual crystal lamp and reasonable match of texture and colors.

2. CONTENT (SOFTNESS)
Style and theme (function: spiritual needs): Sales Place of Wanda has always persisted in cultural property, combined with the local cultural characteristics and uniformed theme and style, in order to satisfy clients' spiritual needs. For example, Dalian Sales Office is finally endowed with the theme of "Ocean Culture" by analyzing and positioning ocean culture of Dalian as a coastal city. Sales place theme will be emphasized by adding accordingly background music and decoration when the place is open to public.

3. ENVIRONMENT (GOOD MOOD)
The excellent atmosphere leaves good impression on the clients, satisfies their purchase desire and finally makes them sign the purchase contract by seeing good view, touching smooth texture, smelling pleasant odor, hearing enjoyable sound and tasting good food. Presentation of a perfect sales place is the effort and result of three departments: Design, Marketing and Construction. Taking a sales place in Zhengzhou as an example, surrounded with healthy and green theme, the sales place has an outdoor garden (green forest is the source of natural oxygen) and an indoor PM2.5 air cleaning system which can both support healthy environment. What is more, different rooms are painted with specific colors, for instance, children room is warm-toned while study room is in cool tone. Combination of different colors and lighting create simple and generous environment in the reception area, easy and lively theme in the kitchen while warm and comfortable atmosphere in the bedroom.

While taking visual impact and use functions as the foundation, the value exploration as the core concept and the atmosphere creating as the guidance, Sales Place of Wanda successfully shows the confidence of Wanda Group as an enterprise with a history of one hundred years and as well as its vision an international company.

3. 有环境（心情好）

看到好看的、触到好质感的、嗅到好闻的、听到好听的、尝到好吃的——氛围营造让潜在客户有代入感，产生共鸣，进而产生购买欲望，最后实现签约。一个完美的销售卖场呈现是设计、营销与施工"三架马车"协力向前的过程与结果。如郑州项目卖场，围绕健康、绿色的主题展开。室外设计了花园（郁郁葱葱的绿色营造天然氧吧），室内设置了PM2.5空气净化系统，支撑一个健康的环境。不同的功能房间采用不同的色调渲染，儿童房采用暖色调，书房采用冷色调。通过不同颜色和灯光的结合，会客区简洁、大方，餐厅厨房轻松明快，卧室温馨安逸（图10）。

（图10）儿童业态样板间效果图

万达卖场坚持以视觉冲击功能及使用功能为基础，以价值挖掘为核心，以氛围营造为导向，诠释万达人百年企业的信心，国际万达的视野。

SHENYANG OLYMPIC WANDA MANSION: SALES OFFICE/ PROTOTYPE ROOM

沈阳奥体万达公馆-售楼处/样板间

开放时间	2012 / 05
建设地点	沈阳 / 辽宁
占地面积	4300 平方米
建筑面积	售楼处: 1150 平方米
	样板间: 730 平方米

OPENED ON	MAY / 2012
LOCATION	SHENYANG / LIAONING PROVINCE
LAND AREA	4,300 m²
FLOOR AREA	SALES OFFICE: 1,150 m²
	PROTOTYPE ROOM: 730 m²

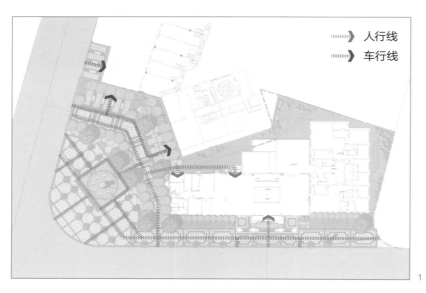

1

2

3

1 沈阳奥体万达公馆售楼处动线图
2 沈阳奥体万达公馆售楼处建筑
3 沈阳奥体万达公馆售楼处夜景

ARCHITECTURE OF THE SALES OFFICE
售楼处建筑

售楼处建筑外装采用古典欧式元素，充分烘托高级住宅和商铺的不凡气质。圆形拱门、挺拔立柱、欧化的重檐线角及凸凹变化外窗等，均是建筑之精彩所在。精雕细琢的夜景照明，更把外立面的特征映衬得淋漓尽致。

The exterior of the sales office applies the classical European-style elements to demonstrate the extraordinary characters of the supreme mansion and shops. The circular arches, straight uprights, European-style double-eave molding, variable convex-concave external windows, etc. are all splendid features of the sales office. Furthermore, delicate design of the night-scape lighting incisively and vividly represents characteristics of the facade in an extreme way.

LANDSCAPE OF THE SALES OFFICE
售楼处景观

设计充分结合建筑的欧式风格，从景观的种植、小品、灯具等方面选用经典造型，烘托出浓郁的异国景观情调。售楼处后的花园，大树环绕，灌木与时花紧密搭配，同时设置了休闲座椅、雕塑、连廊，结合灯光设计，形成一幅室外会客厅的风情画卷。后花园和前广场着重打造营销氛围，通过欧式画廊序列及室外水吧台的展示，对未来住宅环境空间进行浓缩和提炼，再现高品质生活方式。

In combination with European style of architecture, the landscape design of the sales office chooses classic models from landscape plants, ornaments and lamps, in order to create exotic atmosphere. At the back of the sales office, the garden is designed with trees, bushes and seasonal flowers, interspersing with recreational chairs, sculptures and corridors, to form a fabulous scroll painting of an outdoor reception hall under the lighting design background. Focusing on creating marketing atmosphere, the Back Garden and Front Square realize enrichment and extraction of future residential space through European gallery and outdoor water bar counter, and therefore successfully presents a high-quality lifestyle.

4 沈阳奥体万达公馆售楼处连廊
5 沈阳奥体万达公馆售楼处后花园景观
6 沈阳奥体万达公馆售楼处连廊近景

7 沈阳奥体万达公馆售楼处木栈道
8 沈阳奥体万达公馆售楼处花径
9 沈阳奥体万达公馆售楼处室外水吧台

10

INTERIOR DESIGN OF THE SALES OFFICE
售楼处内装

在建筑平面布局上，结合未来商铺的方正性特征，采用欧式传统的平面布局手法，使得整个售楼处空间布局严谨、对称。墙面运用了大量的西班牙米黄石材，简洁又不失典雅高贵。服务台背景墙运用了大花绿石材，使整个空间富于动感，令宾客耳目一新。地面运用了帝王金、棕色的热带雨林和西班牙米黄的石材拼花，使整个大堂更加金碧辉煌，富有层次和气势。

With the consideration of the shape of future stores, the interior design adopts the traditional European plan layout to create rigorous and symmetrical space arrangement of the whole sales office. Wall adopts large quantity of succinct and elegant Cream Marfil marble, and the background wall of reception desk adopts Dark Green marble, creating a dynamic spatial effect and offering guests an entirely new experience. The floor adopts mixed pattern of Gold Imperial, Rainforest Brown and Cream Marfil marble, making the lobby more resplendent and magnificent, grand and better-organized.

11

12

PROTOTYPE ROOM
样板间

餐厅与客厅连成一个空间，宽敞明亮，动线流畅，空间统一性强。客厅墙面造型运用了经典的欧式金线条及造型，彰显气派和风范。天花上的木雕，融入了洛可可元素的精巧与细腻，让人耳目一新。餐厅墙身材料搭配深色木饰面和立体感的墙纸，正中间镶嵌欧式油画，配合顶上华丽的水晶灯，以及华美的餐桌椅和餐具，营造出一种宫廷式的感觉。

The dining room and living room are connected as a whole, creating capacious and bright space with smooth flows and unified spatial effects. Walls of the living room adopt typical European golden moldings, which highlight the gorgeousness and elegance; wood carvings on the ceiling integrate delicateness and exquisiteness of Rococo elements into the design, which could offer clients an entirely new experience; wall materials of the dining room is decorated with dark timber facing and stereoscopic wallpaper, and inlaid with European oil painting, integrated with gorgeous crystal ceiling lamp, fine dining tables and chairs and delicate tableware to create imperial atmosphere.

13

14

15

16

10 沈阳奥体万达公馆售楼处大厅
11 沈阳奥体万达公馆售楼处平面图
12 沈阳奥体万达公馆售楼处装饰图案
13 沈阳奥体万达公馆样板间A客厅
14 沈阳奥体万达公馆样板间A主卧
15 沈阳奥体万达公馆样板间B客厅
16 沈阳奥体万达公馆样板间B主卧

NANNING QINGXIU WANDA MANSION: SALES OFFICE / PROTOTYPE ROOM

南宁青秀万达公馆-售楼处／样板间

开放时间	2013 / 12
建设地点	广西／南宁
占地面积	2500 平方米
建筑面积	1300 平方米
OPENED ON	DECEMBER / 2013
LOCATION	NANNING / GUANGXI ZHUANG AUTONOMOUS REGION
LAND AREA	2,500 m²
FLOOR AREA	1,300 m²

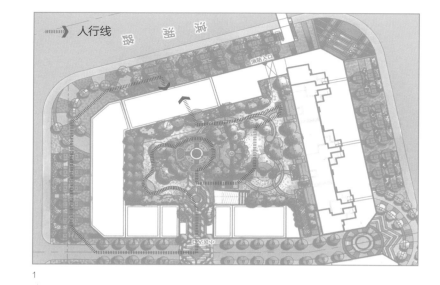

1

2

1 南宁青秀万达公馆动线图
2 南宁青秀万达公馆售楼处大堂
3 南宁青秀万达公馆售楼处休息区

ARCHITECTURE OF THE SALES OFFICE
售楼处建筑

南宁青秀万达公馆售楼处面向城市主干道滨湖路，主体两层，首层为展示洽谈区，二层为员工办公区，占地面积1100平方米，建筑面积1300平方米。沙盘主展示厅直接主入口，挑空两层的200平方米无柱空间尽显尊贵大气。接待区、洽谈区、水吧和影视厅等环绕主展厅布置，方便使用的同时保证各功能分区间的视线交流。入口左侧设公馆实景大堂及实景电梯厅，给予业主直观的入户体验。

Facing Binhu Road, one of the urban main roads in Nanning, the sales office of the Nanning Qingxiu Wanda Mansion is mainly composed of display and negotiation area on the ground floor while staff office area on the first floor. Its land area is 1,100 square meters with a floor area of 1,300 square meters. The main display hall with sand table is directly connected to the main entrance, and the column-free space of the office with two floors of 200 square meters presents dignity and magnificence. The reception area, negotiation area, water bar and movie hall are arranged around the main display hall for the convenience of use and visual interaction between all functional zones. The demonstration lobby and elevator hall, which show live scene of the Mansion, are set on the left side of the entrance, which could provide the owners an intuitive experience.

4 南宁青秀万达公馆入户大堂
5 南宁青秀万达公馆主卧
6 南宁青秀万达公馆客厅
7 南宁青秀万达公馆中央庭院景观
8 南宁青秀万达公馆阳台
9 南宁青秀万达公馆户型图

LOBBY OF THE SALES OFFICE
售楼处大堂

售楼处实景大堂：法式宫廷风格，突出细腻精致的工艺，强调以华丽的装饰、厚重的色彩、精美复古的造型，融合现代元素，达到雍容华贵、细腻雅致的装饰效果。空间特征：布局上强调轴线的对称，沿袭古典欧式的主元素，有机地融入现代的生活元素，散发豪华大气与法式浪漫惬意的空间气息。

The demonstration lobby of the sales office is designed in the French palace style and emphasizes on exquisite workmanship, expecting to create dignified and graceful, delicate and elegant decorative effect by gorgeous ornament, bright color, vintage modeling and modern elements. Spatial layout of the lobby is axially symmetric, and classical European major elements are organically integrated with modern life elements, which created luxury, splendid, romantic and cozy space.

PROTOTYPE ROOM
样板间

主卧室及更衣室地面采用实木复合地板，与其他空间地面材料交接收口部位平接。主卧室设计造型吊顶、暗藏灯带、造型灯盘及古典风格的欧式线角。更衣室吊平顶，欧式阴角线收口，采用暖光源照明。主卧室床背墙做造型，墙面壁纸图案有别于其他部位。窗台板采用大理石材质，石材造型线角做收口处理。

The floors of the master bedroom and dressing room adopt solid wood laminate flooring, while the joints with flooring of other spaces adopt butt joints. The master bedroom is designed with modeling suspended ceiling, concealed strip, modeling lamp panel and the classical European molding. And the dressing room is set with flat suspended ceiling, European internal corner joint and warm illuminant lighting. Modeling bed wall is adopted in the master bedroom, and pattern of its wallpapers is different from that on other places. Window sill is made of marble, and stone modeling molding is blinded off.

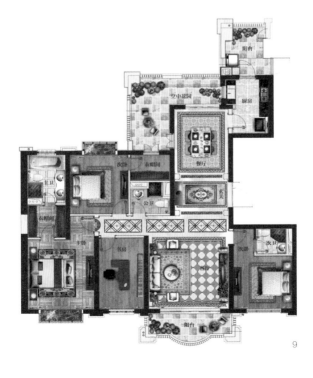

CHANGSHA KAIFU WANDA MANSION: SALES OFFICE/ PROTOTYPE ROOM

长沙开福万达公馆-售楼处/样板间

开放时间	2013 / 11
建设地点	湖南 / 长沙
占地面积	4530 平方米
建筑面积	906 平方米
OPENED ON	NOVEMBER / 2013
LOCATION	CHANGSHA / HUNAN PROVINCE
LAND AREA	4,530 m²
FLOOR AREA	906 m²

1

1 长沙开福万达公馆售楼处后花园
2 长沙开福万达公馆售楼处室外水吧台
3 长沙开福万达公馆售楼处花钵布花方案
4 长沙开福万达公馆"树的故事"主题景观
5 长沙开福万达公馆售楼处竹林夹道

LANDSCAPE OF THE SALES OFFICE
售楼处景观

通过对人的行为心理及营销氛围分析,制定完整的参观动线。前广场以法式古典宫廷景观为主,在主轴线上形成宏大的景观气势——对称的、阵列的、仪式感的景观对景;后花园首次引入室外水吧台和迷你高尔夫的概念,打造欧洲贵族经典的生活场景,对未来住宅环境空间进行浓缩和提炼,再现高品质生活方式。两个空间通过室内及出口的竹林,使空间遵循开一闭一开的节奏,进一步激发消费群体的购买欲望。

A complete visit flow is established on the basis of analysis of human behavior psychology and marketing atmosphere. The front square is mainly arranged with the French classic palace landscapes in symmetrical and array layout of opposite scenery, forming grand and magnificent landscape along the main axis. The back garden introduces outdoor water bar counter and mini-golf for the first time to create a typical scene of aristocratic life in Europe. It also concentrates and extracts housing space in the future, representing high-quality living style. The front square and back garden are designed in an open-enclosed-open layout by taking advantage of the bamboo forest both indoors and at the entrance, in order to further motivate the clients' purchasing desire.

PROTOTYPE ROOM
样板间

设计导入了古典欧式巴洛克风格，运用华丽的装饰、浓烈的色彩，体现出真正高级住宅无处不在的价值和细节魅力，彰显了主人身份、地位及对生活的品味。

The classical Baroque style is applied in the design not only to represent ubiquitous values and charming details of real grade residences, but also to demonstrate distinguished identity and elegant lifestyle of the owner through its unique gorgeous ornaments and bright colors.

9

6　长沙开福万达公馆售楼处迷你高尔夫景观后花园
7　长沙开福万达公馆售楼处竹林夹道
8　长沙开福万达公馆售楼处后花园摆件
9　长沙开福万达公馆样板间主卧化妆台
10　长沙开福万达公馆样板间客厅

10

大堂挑高7米,具有强烈的空间感,配置紫铜门,高贵而典雅。地面是大理石水刀拼花,天花采用金箔贴面,配以施华洛世奇水晶吊灯和艺术挂画,整体风格奢华古典。进入户内,餐厅和客厅通过罗马柱相连,既分隔又相通,宽敞而明亮。门厅地面亦采用图案精美的大理石水刀拼花,与椭圆的天花雕花造型完美呼应。客厅电视墙背景运用了经典欧式拱门的手法,旁边的雕花及装饰柱,尤其是天花上的木雕,精雕细刻,融入洛可可元素的精巧与细腻,整体效果豪华气派。

The lobby is raised for 7 meters, in order to create a strong sense of space; and so does it set with cooper door, to show nobility and elegance of the space. The elements such as matter high-pressure water-jetted marble floor, gold foil veneered ceiling, Swarovski crystal chandelier and the art paintings all represent luxury and classic beauty of the lobby. In the lobby, the dining room is connected with the living room by using the Roman pillars, forming bright and spacious space in a partitioned and interlinked layout. Floor of the hallway also adopts patterned high-pressure water-jetted marble flooring, which is in concert with modeling of the oval ceiling carving. TV wall in the living room is designed in the classical European arch style with elaborate carving patterns, decorative pillars and wood carvings on the suspended ceiling, harmoniously integrating delicateness and exquisiteness of the Rococo elements into the design and creating luxury and gorgeous overall effect.

11 长沙开福万达公馆样板间大堂天花
12 长沙开福万达公馆样板间桌椅
13 长沙开福万达公馆样板间书房

DONGGUAN DONGCHENG WANDA MANSION: SALES OFFICE/ PROTOTYPE ROOM
东莞东城万达公馆-售楼处/样板间

开放时间	2012 / 08
建设地点	广东 / 东莞
占地面积	1800 平方米
建筑面积	1230 平方米
OPENED ON	AUGUST / 2012
LOCATION	DONGGUAN / GUANGDONG PROVINCE
LAND AREA	1,800 m²
FLOOR AREA	1,230 m²

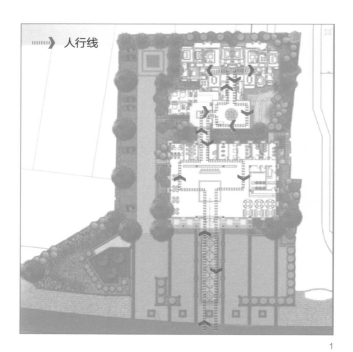

1

ARCHITECTURE OF THE SALES OFFICE
售楼处建筑

该项目销售中心和样板房区建筑立面风格统一，采用追求华贵、高雅的新古典建筑风格。此立面一方面保留了古典欧式建筑中的材质、色彩的大致风格，另一方面可以很强烈地感受传统的历史痕迹与浑厚的文化底蕴，就像欧洲的旧贵族一样，讲究庄重、华贵。设计中着重利用线脚呈现细微而丰富的建筑肌理，窗框、阳台栏杆等采用深色金属构件，结合玻璃、空调机位百叶窗，使建筑因统一的视觉感而更显生动鲜活。

The neoclassical architectural style, which pursues luxury and elegance, is adopted to in the design of the sales office facade and the prototype room of the Dongguan Dongcheng Wanda Mansion. So that texture and color of classical European architecture can be generally reserved, and traditional historical rudiments and rich cultural deposits, such as solemnity, luxury and showy adornments of European aristocracy, can be also fully represented. The design emphasizes on utilizing moldings to present delicate and rich building texture. Metal components in dark colors are adopted for window frames and balcony railings, and other elements, such as glass and air-conditioning louvers, jointly create a unified visual impression and make the building more vivid and lively.

1 东莞东城万达公馆售楼处动线图
2 东莞东城万达公馆售楼处建筑外立面
3 东莞东城万达公馆售楼处入口门厅

PART C WANDA SALES PLACES
万达销售卖场

4 东莞东城万达公馆售楼处大堂

INTERIOR DESIGN OF THE SALES OFFICE
售楼处内装

该项目设计以沉稳豪气的比例构成方式搭配富有经典欧式韵味的线条,尽显欧式皇家之风范;美轮美奂的工艺品以及艺术陈设的运用,在保证空间衔接关系流畅的前提下,追求整体格调的提升与艺术氛围的协调,为空间营造出尊贵气息与典雅的艺术氛围。

In combination with splendid structure and the classical European modeling, the design of the project incisively represents quality of the European royalty, and the fabulous artworks and art display, which pursue both the improvement of the overall style and the coordination of artistic atmosphere on the premise of smooth spatial transition, create noble and elegant artistic atmosphere.

PROTOTYPE ROOM
样板间

该项目的设计灵感源于欧式建筑，室内装饰淋漓尽致地体现出欧式新古典风格。纯粹的大户设计，外观上俊朗的线条尽显力量感。内部配以欧式手工奢装，甄选国际名贵大理石材与世界顶尖家具、电器，处处彰显殿堂级的奢华；但设计希望给予的不仅仅是尊贵，更是一种超越于尊贵之上的生活状态与人生的积累、沉淀。

Design of the prototype room is inspired by European architecture, so the interior design incisively represents the European neoclassical style. This project adopts large house type design. The clear appearance design offers the building a sense of power, and the interior space is decorated in the European luxury style with international well-known marbles and best furniture and electric appliances to display palace-style luxury. However, the design expects to offer the resident not only a kind of honor but also an accumulation and sedimentation of living and life.

7

8

9

5 东莞东城万达公馆入户大堂
6 东莞东城万达公馆客厅
7 东莞东城万达公馆主卧室
8 东莞东城万达公馆卧室陈设
9 东莞东城万达公馆景观小品

YANTAI ZHIFU WANDA MANSION: SALES OFFICE/ PROTOTYPE ROOM
烟台芝罘万达公馆-售楼处/样板间

开放时间	2013 / 07
建设地点	山东 / 烟台
占地面积	5283.9 平方米
建筑面积	990.7 平方米
OPENED ON	JULY / 2013
LOCATION	YANTAI / SHANDONG PROVINCE
LAND AREA	5,283.9 m²
FLOOR AREA	990.7 m²

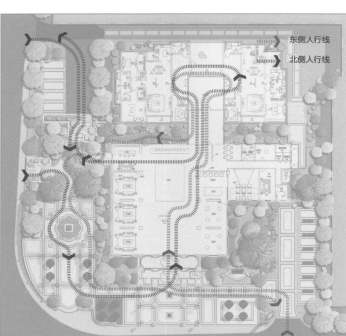

1

ARCHITECTURE OF THE SALES OFFICE
售楼处建筑

烟台芝罘万达公馆售楼处、样板间建筑采用法式建筑风格，三段式的经典设计彰显建筑奢华宏大的气势。建筑造型线条鲜明，石材稳重高贵。建筑室内布局严谨，采用拱顶描金，强化轴线关系，气势恢宏；法式廊柱、雕花、线条等装饰的点缀，很好地诠释了法式建筑的典雅与尊贵。

The sales office and prototype room of the Yantai Zhifu Wanda Mansion are designed in the French architectural style with classic three orders design to present manifests luxury and grand characteristics of the buildings. Modeling lines of the building is clear, and stones appear to be steady and dignified. The indoor layout is orderly with traced arch in gold to highlight axis and strengthen magnificence, integrated with the embellishment of the French columns, craving pattern and lines to well express the elegance and dignity of the French architecture.

1 烟台芝罘万达公馆动线图
2 烟台芝罘万达公馆售楼处夜景

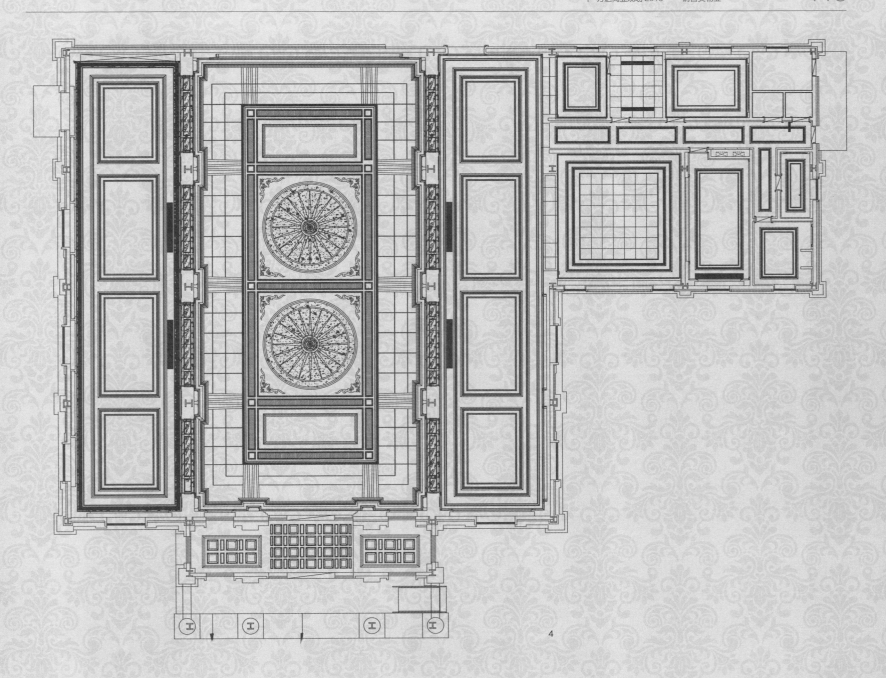

INTERIOR DESIGN OF THE SALES OFFICE
售楼处内装

空间特征强调轴线对称，沿袭古典欧式主元素，有机融入现代生活元素，兼具豪华大气与浪漫惬意的空间感受。

The spatial layout is axially symmetric, and the classic European major elements are organically integrated with modern life elements, providing clients a sense of space of luxury and magnificence, romance and pleasure.

3 烟台芝罘万达公馆售楼处大堂
4 烟台芝罘万达公馆售楼处天花平面图
5 烟台芝罘万达公馆售楼处柱头

WANDA SALES PLACES
万达销售卖场

LANDSCAPE OF THE SALES OFFICE
售楼处景观

景观利用场地沿西南河路方向最大2米高差，通过错台式台阶、阶梯式绿化、多层次植被栽植，营造出趣味十足的看房空间；将销售场地竖向不利转为有利。根据销售场地各方向空间尺度，营造出各具特色的空间序列；小品、雕塑经典并富有情趣，植物丰富的空间层次保证从人的行进道路上360度无死角景观展示，软硬景观结合突出彰显了高级住宅"公馆"的形象。

Taking advantage of the maximum height difference of 2 meters of the site along the Xinanhe Road, landscape of the sales office adopts staggered steps, stepped greening and multilayer vegetation planting to build an attractive show room, changing the vertical constraints into favorable conditions. Various distinctive spatial series are created according to spatial scales of the sales space in each direction. Classic and attractive ornaments and sculptures are scattered in the space, and the spatial pattern with abundant plants can provide 360-degree landscape appreciation for visitors on their way, such combination of soft and hard landscapes further highlight the image of high-end residential Mansion.

6 烟台芝罘万达公馆售楼处到样板间参观通道景观
7 烟台芝罘万达公馆售楼处错台式台阶入口
8 烟台芝罘万达公馆售楼处宝瓶柱围墙

9

10

11

12

9　烟台芝罘万达公馆售楼处前广场情景雕塑
10　烟台芝罘万达公馆售楼处前广场水景夜景
11　烟台芝罘万达公馆售楼处后花园
12　烟台芝罘万达公馆售楼处后花园

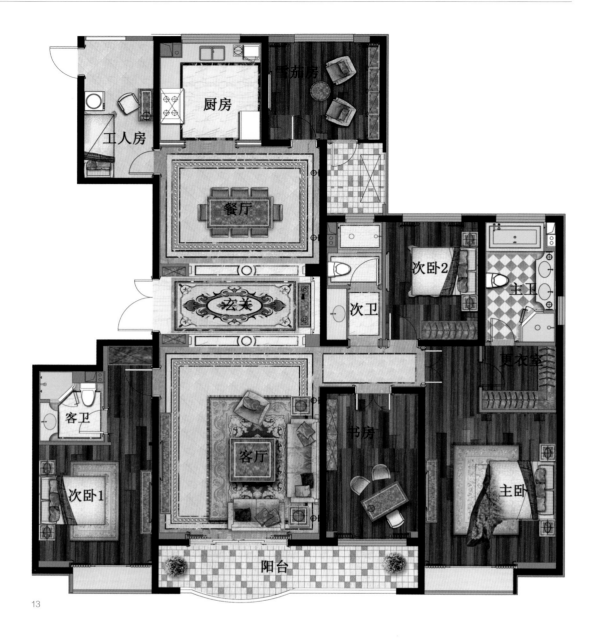

PROTOTYPE ROOM
样板间

样板间工艺精致,还原古典气质,兼具古典、现代双重审美效果,彰显尊贵、奢华、大气的新古典设计风格。色彩搭配以米白色为主基调,辅以素雅花鸟壁纸,简化古典细节装饰。以活跃对比色突出重点部位,整体色调清新自然、细腻雅致。细部造型运用了法式廊柱、雕花、线条,搭配欧式图案。

Exquisite workmanship and classical characteristics of the prototype room express dual aesthetics of classical beauty and modern beauty, and present the dignity, luxury and magnificence of the neoclassical design style. Color design takes creamy white as the key tone, integrating with simple but elegant wallpapers of flower and bird theme and detail decoration of the simplified classical style to highlight key points through color contrast. The overall color design appears to be clean and natural, delicate and elegant. And French columns, craving patterns, lines and complicated European patterns are adopted for detail modeling design.

13 烟台芝罘万达公馆户型图
14 烟台芝罘万达公馆样板间入户大堂

样板间整体效果突出细腻精致的工艺,强调以华丽的装饰、清淡的色彩、精美复古的造型,达到雍容华贵、细腻雅致的装饰效果。细部造型运用了法式雕花、线条,搭配繁复的欧式图案,强调制作工艺上精益求精的细节处理。色彩以米白色为主基调,活跃的对比色突出重点部位,其余部分以浅色调协调统一为原则,整体色彩搭配清新自然。

Overall effect of the prototype room emphasizes exquisite workmanship and expects to create dignified and graceful, delicate and elegant decorative effect by gorgeous ornaments, light colors and fine vintage modeling. Detail modeling adopts French columns, craving patterns, lines and complicated European patterns to emphasize excelsior detail treatment. Color design takes creamy white as the key tone, and key points are highlighted through color contrast, coordinating and integrating with other parts of light color tone. The overall color design appears to be clean and natural.

15 烟台芝罘万达公馆样板间客厅
16 烟台芝罘万达公馆样板间卧室
17 烟台芝罘万达公馆样板间餐厅

NANJING JIANGNING WANDA MANSION: SALES OFFICE

南京江宁万达公馆-售楼处

开放时间	2012 / 04
建设地点	江苏 / 南京
占地面积	1834 平方米
建筑面积	1438 平方米
OPENED ON	APRIL / 2012
LOCATION	NANJING / JIANGSU PROVINCE
LAND AREA	1,834 m²
FLOOR AREA	1,438 m²

INTERIOR DESIGN OF THE SALES OFFICE
售楼处内装

在满足销售功能的基础上，大厅中间部位设置了沙盘展示区和洽谈区。二者相互共享又互不干扰，合理利用了空间。沙盘展示设在中央最显眼的位置；销售人员工作区设在大厅后方。服务台造型高端大气，简单又不失稳重；背景造型突出了售楼部的企业形象与品牌。天花处理上采用了一个整体的半圆弧强调中心主题，加上丰富的色彩变化和描金PU线条的穿插结合，使空间更加丰富多彩。

On the premise of satisfying the sales function, the hall is set with sand table display area and negotiation area in the middle. These two areas are mutually shared but not interfered in reasonable space layout. The sand table is set at the most attractive place in the middle, and working area of sales staff is in the back of the hall. The high-end and splendid service desk appears to be simple and steady, and the background modeling further emphasizes corporate image and brand. An integral semicircle ceiling treatment is adopted to highlight theme, and the combination of color variations and gold tracing lines makes the space more rich and colorful.

1 南京江宁万达公馆动线图
2 南京江宁万达公馆售楼处大堂

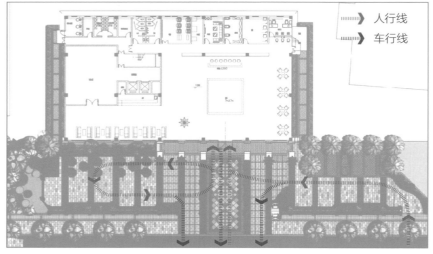

古典欧式风格是一种追求华丽、高雅的建筑风格。在造型上，以欧式线条勾勒出不同的装饰造型，典雅大气。在色彩上，运用明黄、米白等古典风格的常用色来渲染空间氛围，营造出富丽堂皇的效果。屋顶多采用坡屋顶；顶上有精致的老虎窗；外墙用石材装饰，细节处理上运用了法式廊柱、雕花、线条，呈现出浪漫典雅风格。整个建筑采用对称造型，气势恢宏。

The classical European style is a kind of architectural style that pursues for gorgeousness and elegance. European lines are adopted to create various magnificent and elegant decorative modeling, and bright yellow, cream white and other commonly adopted colors in the classical style are utilized to strengthen the spatial atmosphere and build a gorgeous and splendid spatial effect. The roof adopts pitched roof with delicate dormer on top whilst external wall is decorated with stone. The detail treatment well expresses romantic and elegant style through French columns, craving patterns and lines. The whole building adopts symmetrical modeling, appearing to be resplendent and magnificent.

WUHAN K4 WANDA MANSION: SALES OFFICE/ PROTOTYPE ROOM

武汉 K4 万达公馆－售楼处/样板间

开放时间	2013 / 04
建设地点	湖北 / 武汉
占地面积	5600 平方米
建筑面积	2000 平方米
OPENED ON	APRIL / 2013
LOCATION	WUHAN / HUBEI PROVINCE
LAND AREA	5,600 m²
FLOOR AREA	2,000 m²

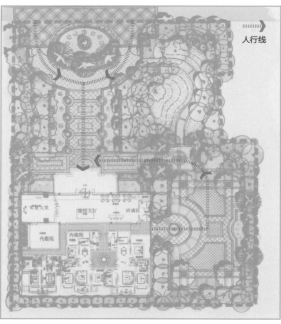

1 武汉K4万达公馆动线图
2 武汉K4万达公馆景观门楼
3 武汉K4万达公馆建筑外立面

ARCHITECTURE OF THE SALES OFFICE
售楼处建筑

武汉K4万达公馆售楼处注重项目本身的连续性，充分利用临水的景观优势，营造良好的日照采光条件。建筑外立面采用欧式线条结合局部收分的手法。项目整体感强烈、轮廓清晰、注重文化和内涵的表达，以及对品质感的极致要求。

The sales office located at the Wuhan K4 Wanda Mansion emphasizes continuity of the project. The landscape advantages given by its waterside location are fully utilized to create favorable sunlight lighting conditions, and the European-style lines in combination with partial batters are applied in the facade design. The project enjoys fine wholeness and clear profile, emphasizes expression of culture and connotation and pursues perfect quality experience.

5

PROTOTYPE ROOM
样板间

采用法国古典风格——华丽雕琢、纤巧精致的工艺结合优雅古典元素——映射18世纪法式尊贵、高雅的古典美。大堂给人以华美、考究及富于文化内涵的感觉。墙面为18世纪传统法式米色调的干挂石材，以及高大的香槟红石材双柱，配搭18世纪油画，为大堂空间平添尊贵与气派。

The prototype room is designed in the French classical style, which is expressed by gorgeous carving, exquisite workmanship and elegant elements, reflecting French classical beauty of dignity and elegance of the eighteenth century. The lobby appears to be gorgeous, exquisite and cultured, where traditional French cream dry-hung stones of the eighteenth century are adopted as wall materials and tall paired columns of the Perlino Rosato stone, embellishing with oil paintings of eighteenth century style, making the lobby more dignified and gorgeous.

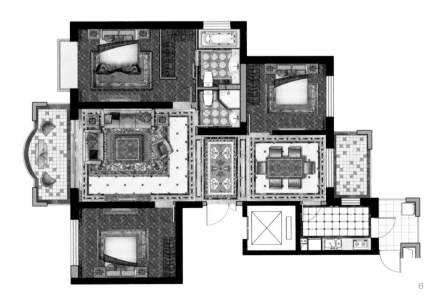

6

4　武汉K4万达公馆样板间入户大堂
5　武汉K4万达公馆样板间客厅
6　武汉K4万达公馆样板间B户型图

样板间采用部分法式古典装饰风格的花鸟主题元素，华丽雕琢的艺术样式结合得古典优雅。各空间均体现法式贵族特有高雅与浪漫的品质。

The prototype room adopts decorative elements of flower and birds theme of the French classical style, integrating with exquisite carving and classical elegance to enable each space to present the quality of unique grace and peculiar romance of French nobility.

8

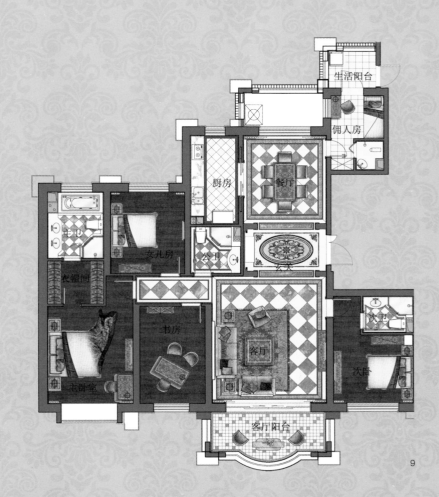

9

7 武汉K4万达公馆样板间客厅
8 武汉K4万达公馆样板间主卧
9 武汉K4万达公馆样板间C户型图

XI'AN DAMINGGONG WANDA MANSION: SALES OFFICE/ PROTOTYPE ROOM

西安大明宫万达公馆-售楼处/样板间

开放时间	2012 / 04
建设地点	陕西 / 西安
占地面积	7118 平方米
建筑面积	2321 平方米
OPENED ON	APRIL / 2012
LOCATION	XI'AN / SHAANXI PROVINCE
LAND AREA	7,118 m²
FLOOR AREA	2,321 m²

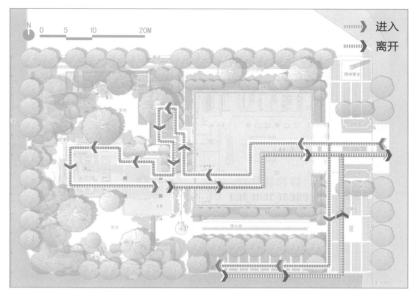

1 西安大明宫万达公馆售楼处外立面
2 西安大明宫万达公馆售楼处动线图
3 西安大明宫万达公馆售楼处景观

ARCHITECTURE OF THE SALES OFFICE
售楼处建筑

西安大明宫万达公馆售楼处建筑采用法式建筑风格，建筑造型线条鲜明，凹凸有致，外观造型独特，颜色稳重大气，呈现出一种华贵之气。建筑室内布局严谨，突出轴线的关系和恢宏的气势，满足基本使用功能的同时彰显其艺术魅力。法式廊柱、雕花、线条等装饰的点缀更是增添细节，细腻地表现了法式建筑的古典与尊贵。

The sales office of Xi'an Daminggong Wanda Mansion is designed in the French architectural style. The clear and distinct architectural modeling, unique appearance and steady color design endow the building with luxury and magnificence. Rigorous indoor layout stresses the axial arrangement and magnificent momentum and presents the artistic charm on the premise of satisfying basic functions. Decorative ornaments, such as the French columns, craving patterns and lines, enrich the detail of the building and further express the elegance and dignity of the French architecture.

INTERIOR DESIGN OF THE SALES OFFICE
售楼处内装

西安大明宫万达公馆售楼部内装布局上以轴线的对称为主，凸显恢宏的气势和豪华舒适的空间感。细部造型上，运用了法式廊柱、雕花、线条、金箔，搭配以细腻的欧式图案，强调制作工艺精细考究；整体效果华美、大气，呈现美式田园的恬静与法式宫廷的气势，使客户体验成功者的愉悦。

Interior design layout of the sales office of the Xi'an Daminggong Wanda Mansion is generally axially symmetrical, highlighting the magnificent momentum and luxury and comfortable spatial experience. For detail treatment, French columns, craving patterns, lines, gold foil and complicated The European patterns are flexibly utilized, emphasizing the excelsior processes. The overall effect appears to be gorgeous and magnificent, expressing quietness and peacefulness of the American country style and magnificence of the French palace style, bringing clients pleasant experience of the success.

4 西安大明宫万达公馆售楼处大厅

5

6

PROTOTYPE ROOM
样板间

室内布局上强调轴线的对称，细部造型运用了法式廊柱、雕花、线条、金箔，搭配丰富的欧式图案，强调制作工艺精细考究；整体效果呈现美式恬静与法式宫廷皇家气势相结合的风格，突出高贵优雅的气质和细腻精致的工艺。色彩搭配以米色为主基调，辅以香槟金色细节装饰，重点部位以活跃的对比色突出，其余部分以浅色调协调统一为原则，整体色彩搭配突出清新自然的特点。

The interior layout of the prototype room emphasizes on axial symmetry. Detail modeling adopts French columns, craving patterns, lines, gold foil and elaborates the European patterns with excelsior processes, and the overall effect reflects a combination of quietness and peacefulness of the American country style and magnificence of the French palace style, highlighting the nobility and elegance and delicate processes. Color design takes creamy white as the key tone and Champaign gold details as supplementary. Key points are highlighted through color contrast, and other parts adopt light color for coordination and uniformity. The color design appears to be clean and natural in general.

5 西安大明宫万达公馆样板间客厅
6 西安大明宫万达公馆样板间餐厅
7 西安大明宫万达公馆样板间客厅
8 西安大明宫万达公馆样板间客厅
9 西安大明宫万达公馆样板间主卧

ZHENGZHOU JINSHUI WANDA MANSION: SALES OFFICE
郑州金水万达公馆-售楼处

开放时间	2013 / 12
建设地点	河南 / 郑州
占地面积	2037 平方米
建筑面积	2458 万平方米
OPENED ON	DECEMBER / 2013
LOCATION	ZHENGZHOU / HENAN PROVINCE
LAND AREA	2,037 m²
FLOOR AREA	2,458 m²

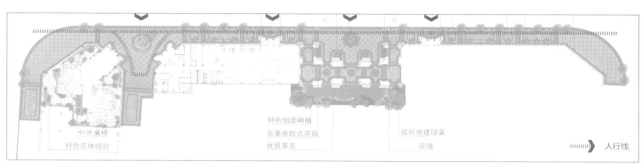

ARCHITECTURE OF THE SALES OFFICE
售楼处建筑

售楼处建筑采用经典欧式造型，立面设计力求高贵典雅，同时反映出建筑坚固有力的风格特征；另一方面，通过稳重的卡拉麦里金石材和精致的细节设计，体现奢华、大气的品质感。

The sales office adopts the classical European modeling, and facade design strives for nobility and elegance on the premise of reflecting firm and solid characteristics of the building. Additionally, steady Karamori Gold stones and delicate detail design display the luxury and magnificent quality.

1 郑州金水万达公馆建筑外立面
2 郑州金水万达公馆动线图
3 郑州金水万达公馆主入口大门

4

LANDSCAPE OF THE SALES OFFICE
售楼处景观

售楼处景观处处体现了浪漫、高贵的情结，体现"艺术与人文、尊贵与成就、古典与现代完美融合的新古典欧式庭院"设计理念。用雕塑喷泉、具有仪式感的五头灯柱、精致的花瓶以及自然柔和的植物组团等作为庭院中的主要元素，以一种尊贵的气派、温文尔雅的姿态呈现经典欧式艺术景观，将雍容华贵的气质尽展无遗。考虑对景需要，内庭中心布置端景内庭。

The landscape of the sales office incisively expresses the romance and dignity and reflects the design concept of the "Neoclassical European Courtyard Style of Combination of Arts and Humanities, Honor and Achievement, Classical and Modern". Main elements of the courtyard, such as fountain with sculptures, superb five-head lampposts, delicate vases and natural plant groups, display the classical European art landscape in a dignified and gentle mode, and fully expresses dignified and graceful quality of the building. In consideration of opposite scenery design, side view inner courtyard is set in the centre of the inner courtyard.

5

INTERIOR DESIGN OF THE SALES OFFICE
售楼处内装

售楼处室内装修整体彰显新古典主义风格的磅礴气势和精美氛围，大厅中央绚彩的造型顶、约9米的挑空、造型顶中部突出部分做泛光照明，演绎出壮美的幻化灯光美景；地面采用大理石水刀拼花，拼花图案古典奢华，线条流畅，色彩搭配沉稳丰富，突出空间的中心感。天花设计造型吊顶、暗藏灯带，立体感强，天花角线选择古典欧式纹样。采用天然大理石柱，柱身形式严谨、比例完整，提升空间的高贵气质。

The Interior design of the sales office reflects the magnificence and delicateness of the neoclassical style. Gorgeous suspended ceiling in the middle of the lobby, about raised space of 9 meters and protruded parts of flood lighting in the centre of the suspended ceiling create a fantastic lighting effect. The high-pressure water-jetted marble floor, classical and luxury patterns, smooth shapes and steady color design all highlight a sense of centre of the space. The modeling suspended ceiling, concealed strip and the classical European molding make the space more vivid and dynamic. Natural marble pillars in formal style and complete scale greatly improve dignity of the space.

4 郑州金水万达公馆售楼处内花园
5 郑州金水万达公馆金街入口处喷泉
6 郑州金水万达公馆售楼处接待台
7 郑州金水万达公馆售楼处屋顶

DONGGUAN HOUJIE WANDA PALACE: SALES OFFICE

东莞厚街万达华府－售楼处

开放时间	2013 / 12
建设地点	广东 / 东莞
占地面积	5300 平方米
建筑面积	1070 平方米
OPENED ON	DECEMBER / 2013
LOCATION	DONGGUAN / GUANGDONG PROVINCE
LAND AREA	5,300 m²
FLOOR AREA	1,070 m²

1 东莞厚街万达华府售楼处建筑外立面
2 东莞厚街万达华府售楼处立面图
3 东莞厚街万达华府售楼处动线图

1

ARCHITECTURE OF THE SALES OFFICE
售楼处建筑

售楼处建筑立面采用新古典主义风格演绎传统文化中的精髓，不仅拥有典雅、端庄的气质，并体现突出的时代特征，散发浓郁的历史痕迹与浑厚文化底蕴，同时又摒弃了过于复杂的肌理和装饰，简化了线条，将怀古的浪漫情怀与现代人对生活的需求相结合，兼具华贵典雅、时尚现代的风格，反映个性化的美学观念和文化品位。

The facade of the sales office is designed in the neoclassical style to express essence of traditional culture. The design presents not only elegant and dignified quality, but also highlights characteristics of the times; abandons the excessively complicated textures and ornaments, simplifies the lines and displays rich historical scent and cultural deposits. The design perfectly combines the nostalgic romantic theme with modern life demands to reflect personalized aesthetic conception and cultural taste with senses of elegance, fashion and modernity.

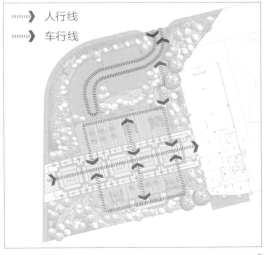

4

5

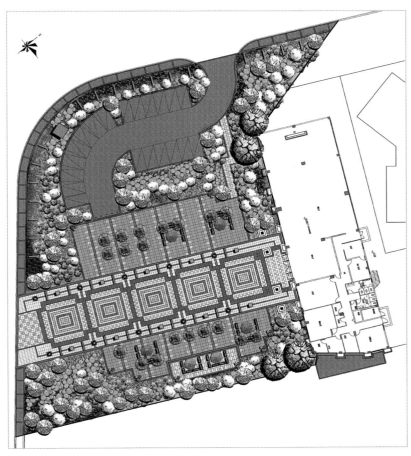

LANDSCAPE OF THE SALES OFFICE
售楼处景观

以简洁的现代欧式设计手法对售楼处前广场功能分区进行合理组织；主通道两侧的景观灯柱、花钵及广告道旗形成强烈的序列感，强化了中央通道，体现尊贵的商业空间感；并在需要进行营销展示等大型活动时，可以灵活有效地利用广场空间。

The front square of the sales office is rationally organized with different function zones in the modern European style. Landscape lampposts, feature pots and advertising boards form an intensive sense of order, strengthening visual effect of central pathway and reflecting a spatial sense of commerce. Such design is favorable for effective utilization of the square for marketing fair or other large-scale events.

4 东莞厚街万达华府售楼处前广场
5 东莞厚街万达华府售楼处景观布置
6 东莞厚街万达华府售楼处景观总平面图
7 东莞厚街万达华府售楼处花钵

INTERIOR DESIGN OF THE SALES OFFICE
售楼处内装

设计上重视形式构图方面的美感，采用对称、结构变换、重复组合等多种手法营造出典雅、自然高贵的气质。以浅色调为基调，深色调为辅贯穿于空间中，色彩的运用给人一种高端大气、温和安静的居室感觉。

The interior design emphasizes the beauty and aesthetics in form and structure through symmetric design, structure change, repetition and combination, and other methods to create an elegant and dignified image. The color design of the sales office takes light color as the basic tone, assisting by dark color, offering people magnificent and comfortable living experience.

8

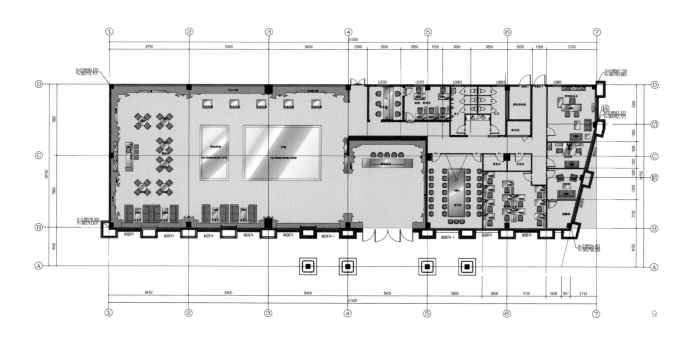

8 东莞厚街万达华府样板间入户大堂
9 东莞厚街万达华府售楼处平面图
10 东莞厚街万达华府售楼处连廊
11 东莞厚街万达华府售楼处大堂

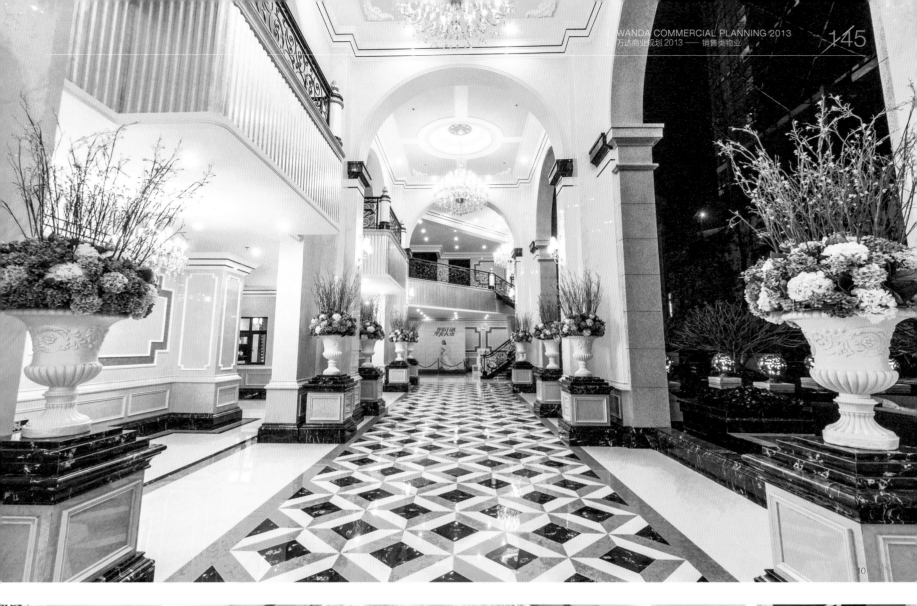

DEZHOU WANDA PALACE: SALES OFFICE/ PROTOTYPE ROOM
德州万达华府-售楼处/样板间

开放时间	2013 / 09
建设地点	山东 / 德州
占地面积	2640 平方米
建筑面积	2148.9 平方米
OPENED ON	SEPTEMBER / 2013
LOCATION	DEZHOU / SHANDONG PROVINCE
LAND AREA	2,640 m²
FLOOR AREA	2,148.9 m²

ARCHITECTURE OF THE SALES OFFICE
售楼处建筑

售楼处采用新古典风格,利用北侧商铺作为样板间,南侧商铺作为售楼处及商铺展示区,中部为Art-Deco风格的入口大门。三者沿街"一"字分布,采用全石材立面,打造精致、尊贵、优雅的立面风格。项目临街面主要为石材幕墙,采用荔枝面黄金麻干挂石材,局部为仿石涂料;玻璃幕墙位于售楼处及样板间入口处,外饰仿铜色金属花饰,体现古典建筑的细腻和尊贵感。

The sales office is designed in neoclassical style. Stores on the north side is arranged as prototype rooms while stores on the south side is set as the sales office and display area with entrance gate of the Art-Deco style in the center. These three sections are distributed longitudinally along the street with stone facade to show delicate, dignified and elegant design style. Frontages of the building are mainly stone curtain walls, where bush-hammered golden grain dry-hung stones and stone like coating are applied. Glass curtain walls coated with bronze metal ornaments are located at entrance of the sales office and prototype room to display delicateness and dignity of the classical architecture.

1

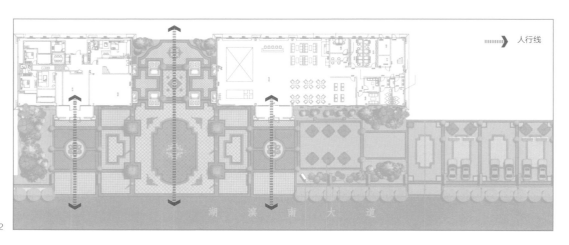

2

1 德州万达华府售楼处大门
2 德州万达华府售楼处动线图
3 德州万达华府售楼处立面图
4 德州万达华府售楼处景观

PART C | WANDA SALES PLACES
万达销售卖场

5 德州万达华府售楼处大堂
6 德州万达华府样板间大堂
7 德州万达华府售楼处室内平面图

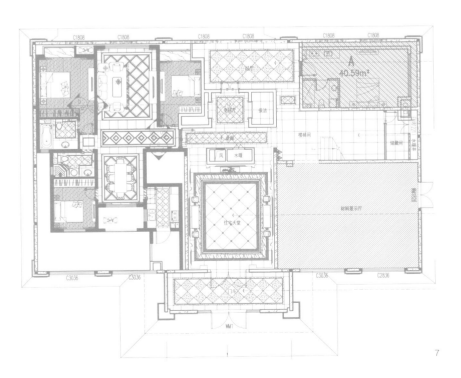

INTERIOR DESIGN OF THE SALES OFFICE
售楼处内装

售楼处的内装，结合了城市整体的经济、文化、审美水平，最终选用欧式风格。设计过程中进行有针对性的细化调整，增加香槟金箔、玫瑰金镜面不锈钢和蚀花银镜等高档面材的使用；增加墙面造型细节，合理把控尺度比例，使造型挺拔大气又不失精巧细致。主形象墙采用名贵松香玉石材搭配背发光拉丝铜logo，营造非凡气度。立体感突出的地面铺装、通透灵动的顶面造型、大气豪华的水晶吊灯，带来第一眼的强烈震撼，充分体现万达大气、尊贵的领袖气质。

Based on the study of integral economic, cultural and aesthetical levels of Dezhou city, the interior design adopts the European style with refined adjustments, like adding Champaign gold foil, rose gold stainless steel with mirror finish, etched silver mirror and other quality plane materials. The designed details of wall modeling make the building towering, magnificent and delicate through appropriate dimension and scale. The main image wall is made of rare Resin Onyx marble with back luminous brushed cooper logo, appearing to be extraordinary and majestic. Stereoscopic floor pavement, dynamic ceiling modeling and luxury crystal chandelier impress clients at the first glance, which fully represents magnificence and dignity of the bellwether quality of Wanda Group.

PROTOTYPE ROOM
样板间

样板间为高贵气质和浪漫风情的法式风格，在材料选择、施工工艺上进行优化，提升了品质及业主感受。人性化设计，打造温馨小两居、甜蜜小三居、幸福大四居；以经典的建筑风格、成熟的生活品质与稀缺的人文资源，构成德州住宅文化的新标杆。

The prototype room is designed in the elegant and romantic French style. Optimized material selection and construction process improve the quality and clients' satisfaction. Humanized design provides different house models, such as warm two-bedroom model, sweet three-bedroom model and blissful four-bedroom model. This project, with the classic architectural style, mature life quality and rare humanistic resources becomes a benchmark of residential culture in Dezhou, Shandong Province.

11

12

13

8 德州万达华府样板间客厅
9 德州万达华府样板间主卧
10 德州万达华府样板间次卧
11 德州万达华府样板间小摆件
12 德州万达华府样板间小摆件
13 德州万达华府样板间小摆件

ZHEJIANG JIAXING WANDA PALACE: PROTOTYPE ROOM
浙江嘉兴万达华府-样板间

开放时间	2013 / 11
建设地点	浙江 / 嘉兴
占地面积	675 平方米
建筑面积	984 平方米
OPENED ON	NOVEMBER / 2013
LOCATION	JIAXING / ZHEJIANG PROVINCE
LAND AREA	675 m²
FLOOR AREA	984 m²

1　浙江嘉兴万达华府样板间入户大厅
2　浙江嘉兴万达华府样板间85平方米样板间客厅
3　浙江嘉兴万达华府样板间115平方米样板间户型图

85m² PROTOTYPE ROOM
85 平方米样板间

85平方米样板间空间装饰采用简洁、硬朗的直线条，搭配中式风格。直线装饰在空间中的使用，不仅反映出现代人追求简单生活的居住要求，更迎合了中式家具追求内敛、质朴的设计风格，使"新中式"更加实用、更富现代感。同时又打破了传统中式空间布局中等级、尊卑等文化思想的界定，空间配色上也更为轻松自然。

The prototype room is designed in the Chinese style with simple and straight finishing lines to not only satisfy the living demands of pursuing simple life of modern people and also cater the Chinese style of modesty and humility, making the "Neo-Chinese" style more practical and modern. Meanwhile, constraints of different classes in the traditional Chinese layout are abandoned and the color design is in a more flexible and natural way.

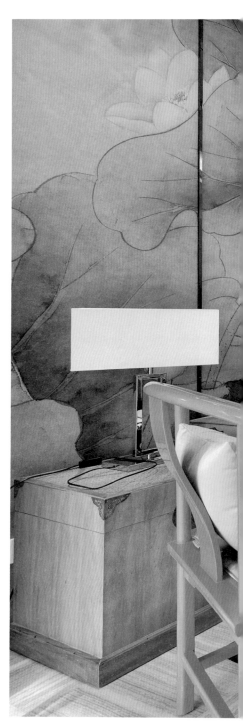

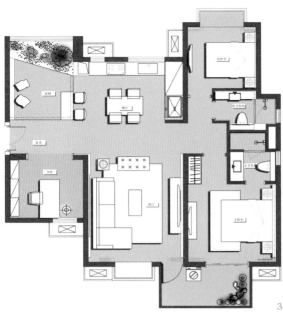

140m² PROTOTYPE ROOM
140 平方米样板间

华丽的装饰、浓烈的色彩、精美的造型达到雍容华贵的装饰效果。顶部大型灯池，用带有花纹的石膏线勾边，并用华丽的枝形吊灯营造气氛。用现代手法还原古典气质，更像是一种多元化的思考方式，将怀古的浪漫情怀与现代人对生活的需求相结合，兼容华贵典雅与时尚现代。家具的主要特色是强调力度、变化和动感；沙发华丽的布面与精致的雕刻搭配，把高贵的造型与地面铺饰融为一体。

Gorgeous decoration, strong color and delicate modeling create a dignified and graceful finishing effect. Large-scale lighting sets on the ceiling are edged with decorative plaster molding and gorgeous chandelier is installed to create atmosphere. Expressing the classical qualities in a modern way is more like a kind of diversified thinking mode, which combines the nostalgic and romantic theme with modern life demands to reflect senses of elegance and modernity. Furniture emphasizes texture, change and dynamic. The sofa with luxury cloth and delicate carving nicely integrate the magnificent modeling with floor pavement.

4 浙江嘉兴万达华府样板间140平方米样板间户型客厅
5 浙江嘉兴万达华府样板间140平方米样板间户型图
6 浙江嘉兴万达华府样板间115平方米样板间餐厅
7 浙江嘉兴万达华府样板间115平方米样板间主卧

JIAMUSI WANDA PALACE: SALES OFFICE
佳木斯万达华府-售楼处

开放时间	2013 / 07
建设地点	黑龙江 / 佳木斯
占地面积	4000 平方米
建筑面积	2300 平方米
OPENED ON	JULY / 2013
LOCATION	JIAMUSI / HEILONGJIANG PROVINCE
LAND AREA	4,000 m²
FLOOR AREA	2,300 m²

1 佳木斯万达华府售楼处动线图
2 佳木斯万达华府售楼处立面图
3 佳木斯万达华府售楼处夜景

ARCHITECTURE OF THE SALES OFFICE
售楼处建筑

售楼处建筑为新古典风格，在传统美学的范畴之下，运用现代的材质及工艺，演绎传统欧洲文化中的经典，不仅拥有典雅、端庄的气质，并具有明显的时代特征。

The sales office is designed in the neoclassical style to express essence of the European traditional culture by modern materials and processes within traditional aesthetic scope. The design presents not only elegant and dignified quality, but also prominent characteristics of the times.

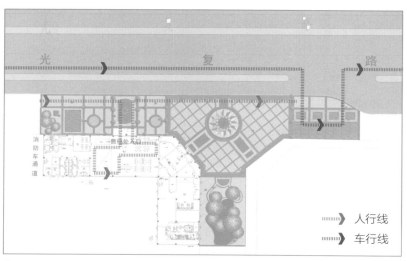

人行线
车行线

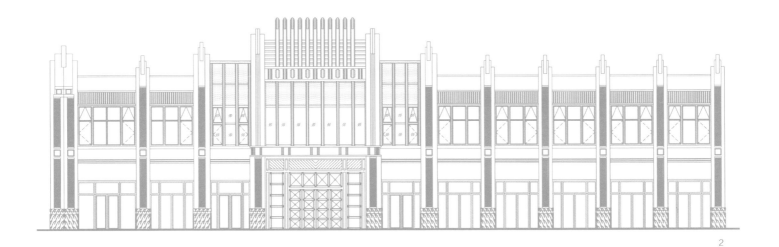

4 佳木斯万达华府售楼处大堂
5 佳木斯万达华府样板间主卧
6 佳木斯万达华府样板间主卧
7 佳木斯万达华府样板间客厅

PROTOTYPE ROOM
样板间

样板间根据新古典风格采用现代的手法和材质,参照古典主义的文化底蕴、美感及艺术气息,将繁复的家居装饰凝练得更为简洁精雅,为生硬的线条配上温婉雅致的软性装饰,将古典美注入简洁实用的现代设计中,使得家居装饰更有灵性。卧室以匀称的线条、金银调的色彩,以及低调奢华的细节而引人入胜。硬装与软装结合得恰到好处。新古典主义的装修风格,让人们体会到古典的优雅,并尽享尊贵雍容。

Referring to the cultural deposits, aesthetics and autistic of classicism, the design of the prototype room, which adopts modern methods and materials in the neoclassical style, refines complicated home furnishings to be more concise and elegant, endows rigid lines with flexible and graceful soft ornaments, in an attempt to integrate classical beauty with the modern design, which makes home furnishings more flexible and ideal. The neat lines, gold-and-silver colors and details of modest luxury make the bedroom more attractive. Hard and soft decorations are well combined in the design, and the neoclassical finishing style brings people elegant, graceful and dignified living experience.

PART C

WANDA SALES PLACES
万达销售卖场

MIANYANG CBD WANDA PALACE: PROTOTYPE ROOM
绵阳 CBD 万达华府－样板间

开放时间	2013 / 04
建设地点	四川 / 绵阳
占地面积	1700 平方米
建筑面积	1892 平方米
OPENED ON	APRIL / 2013
LOCATION	MIANYANG / SICHUAN PROVINCE
LAND AREA	1,700 m²
FLOOR AREA	1,892 m²

85m² PROTOTYPE ROOM
85 平方米样板间

85平方米样板间以法式经典为主调，色调为蓝白。新古典家具与设计氛围的结合，使空间呈现高贵时尚和具备艺术气息的双重品质，体现男女主人年轻、时尚、艺术化的审美理念。

The prototype room is designed in the French classic style with blue and white colors. Matching of the neoclassical furniture and design atmosphere make the space present dual qualities of noble fashion and artistic scent, reflecting aesthetic concept of young, fashion and art of clients.

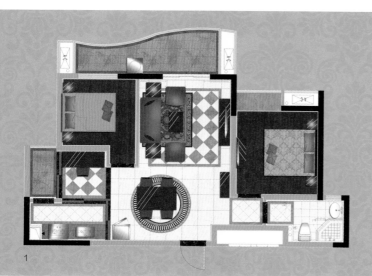

1

1 绵阳CBD万达华府85平方米样板间户型图
2 绵阳CBD万达华府85平方米样板间客厅
3 绵阳CBD万达华府85平方米样板间主卧

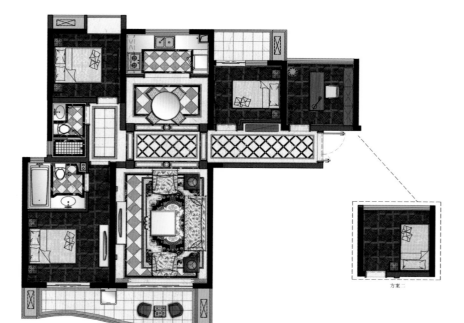

125m² PROTOTYPE ROOM
125 平方米样板间

125平方米样板间以经典青花瓷为设计主调，属于自然风格的一支，倡导"回归自然"，在室内环境中力求表现悠闲、舒畅、自然的田园生活情趣；巧妙地运用了天然木、石、藤等材质质朴的纹理；精巧地设置室内绿化，创造自然、简朴、高雅的氛围。

The prototype room is designed in classical blue and the white porcelain style, which belongs to the natural style and advocates "Back to Nature", in an attempt to display an easy, leisurely and natural life style. Patterns of natural wood, stone, rattan and other simple materials are flexibly used in the design, and indoor greenings elaborately creates natural, unadorned and elegant atmosphere.

4 绵阳CBD万达华府125平方米样板间客厅
5 绵阳CBD万达华府125平方米样板间户型图
6 绵阳CBD万达华府125平方米样板间餐厅
7 绵阳CBD万达华府125平方米样板间主卧

WEINAN WANDA PALACE: SALES OFFICE/ PROTOTYPE ROOM

渭南万达华府－售楼处／样板间

开放时间	2013 / 08
建设地点	陕西 / 渭南
占地面积	823.77 平方米
建筑面积	1022 平方米
OPENED ON	AUGUST / 2013
LOCATION	WEINAN / SHAANXI PROVINCE
LAND AREA	823.77 m²
FLOOR AREA	1,022 m²

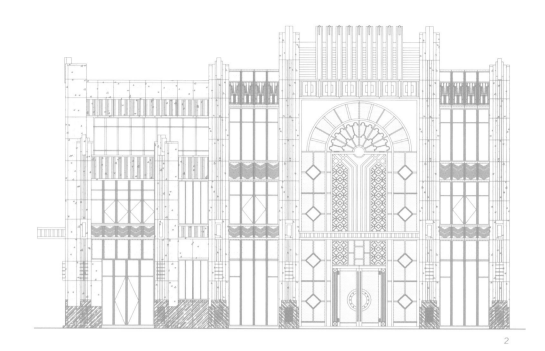

ARCHITECTURE OF THE SALES OFFICE
售楼处建筑

售楼处采用法式建筑风格，外观造型独特，颜色稳重大气，呈现出华贵大气的建筑外观。建筑室内布局严谨，轴线关系明确，装饰层次分明，颜色搭配鲜明，气势恢宏，在满足基本使用功能的同时彰显其艺术魅力，法式雕花、线条等装饰的点缀更是彰显华贵，很好地体现了法式建筑的古典与尊贵。

The sales office is designed in the French architectural style and enjoys distinctive modeling and steady color, appearing to be gorgeous and magnificent. Rigorous indoor layout with clear axial arrangement, and well-arranged decorative layers and distinct color matching present artistic charm with imposing momentum on the premise of satisfying basic functions. French columns, lines and other decorative embellishments well express the elegance and dignity of the French architecture.

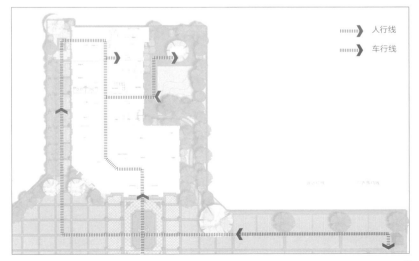

1 渭南万达华府售楼处夜景
2 渭南万达华府售楼处立面图
3 渭南万达华府售楼处动线图

LANDSCAPE OF THE SALES OFFICE
售楼处景观

根据场地特点，通过对植物的高矮、品种搭配、小品元素、硬质铺装等方面的设计，以不同的景观元素营造良好的空间环境和曲径通幽的空间感；以高贵典雅的五头灯配合精美的浮雕底座，排列在卖场通道两侧，取得装饰与功能的统一。情景体验区设计采用一收一放的设计手法，使有限的空间得到有效的放大，通过蜿蜒的园间小路进入花团锦簇的开放空间，营造万达项目的生活品质及生活方式。

Basing on characteristics of the site, the landscape design of the sales office takes advantages of various landscape elements and creates well and quite space environment through different planting layers and assortments, ornaments and rigid pavement, etc. Superb five-head lamps with delicate embossed holder are arranged on both sides of the passageway to achieve decorative and functional uniformity. The scene experience area is designed in a loose and compacted combination mode to effectively enlarge the limited space. The winding path leading to an ornamental open space reflects life quality and living style of Wanda project.

4 渭南万达华府售楼处景观
5 渭南万达华府售楼处景观花钵
6 渭南万达华府售楼处夜景泛光
7 渭南万达华府售楼处景观夜景

灯光通过对建筑设计的理解,渲染营造表现出高品质的视觉效果。夜晚的建筑在灯光的衬托下,充分表达其"雕塑"个性特征——庄重、简约、具有雕塑感,彰显建筑气质,突显建筑特征。

In combination with understandings on the architectural design, lighting design of the project creates a high-quality visual effect, and incisively expresses solemn and concise characteristics of the "sculptural" building under which the lighting at night, highlighting the architectural quality and features.

PROTOTYPE ROOM
样板间

办公室功能样板间，现代设计中融入中式设计元素，调制出时尚温馨的情调；从地面延续到所有主要装饰细节的木质，与浅色恰成对比，而且增添了温暖的气息。在这块所谓中性的"画布"上，黑色元素的介入，加强了基础色彩布局，并凸显了空间之内的独特性。

With function of office, the prototype room integrates modern design with Chinese design elements to create fashionable and warm atmosphere. Wood used by floors and every main decorative detail forms a strong color contrast with the light color and strengthens the warm atmosphere. This prototype room, the neutral "Painting Canvas", introduces black elements into the design to enrich the basic color scheme and emphasize the unique interior design of the space.

8

9

摄影棚风格样板间，结合摄影工作室的工作特点和业务流程，合理区分功能空间，主色调采用淡雅咖啡色调，配上展示性很强烈的欧式家具及其他道具，满足客户在观看体验中的需求和感官体验。

The prototype room takes operating features and business process of photo studio into consideration and rationally separates functional areas. The room is mainly designed in elegant brown and set with the European style furniture and other properties, which are suited for exhibitions, in an attempt to satisfy requirements of clients on viewing experience and sensory experience.

10

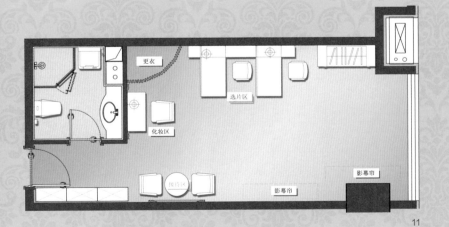
11

12

8 渭南万达华府样板间软装配饰
9 渭南万达华府办公室功能样板间
10 渭南万达华府摄影棚功能样板间
11 渭南万达华府摄影棚户型图
12 渭南万达华府摄影棚功能样板间化妆台

FOSHAN NANHAI WANDA PALACE: PROTOTYPE ROOM
佛山南海万达华府－样板间

开放时间	2013 / 09
建设地点	广东 / 佛山
占地面积	1140 平方米
建筑面积	938 平方米
OPENED ON	SEPTEMBER / 2013
LOCATION	FOSHAN / GUANGDONG PROVINCE
LAND AREA	1,140 m²
FLOOR AREA	938 m²

1 佛山南海万达华府客厅
2 佛山南海万达华府入户大堂
3 佛山南海万达华府餐厅
4 佛山南海万达华府入户花园

1

PROTOTYPE ROOM
样板间

佛山南海万达华府住宅样板房分为三种户型：85平方米现代风格、125平方米港式奢华风格、144平方米新古典主义风格。整个设计重点强调空间与生活之间的紧密依存，以扩大空间及使用机能作为设计主线，强调不夸耀的整体设计与对精致细节的重现；大量使用镜与不锈钢的穿插与点缀，配合高品质墙纸、仿皮，运用视觉上的软硬冲击，延展了多样化的单元空间。

The prototype room of the Foshan Nanhai Wanda Palace covers three house types: the modern style apartment of 85 square meters, the Hong Kong-style luxury apartment of 125 square and the neoclassical style apartment of 144 square meters. The design emphasizes the connection between space and life and the aim of enlarging space and enriching functions, presenting unpretentious overall design and recurrence of delicate details. Mirror and stainless steel materials are largely used in an alternate and interspersed way, coordinating with high-quality wallpapers, leather papers, and visual impact of different textures enlarge the diversified unit space.

2

3

4

5 佛山南海万达华府样板间客厅
6 佛山南海万达华府样板间客厅
7 佛山南海万达华府样板间主卧

材料上点缀镜钢及使用高品质墙纸、皮革等装饰材料，突出细节品质，又更适合现代生活对休闲和舒适的要求；彰显多元化的空间享受，营造出高品质生活氛围。

Mirror and stainless steel materials, high quality wallpapers and leather papers highlight the quality of detail on the premise of satisfying requirements of modern life on relaxation and comfort, displaying diversified space experiences and creating high-quality living environment.

JINING WANDA PALACE: PROTOTYPE ROOM
济宁万达华府－样板间

开放时间	2013 / 09
建设地点	山东 / 济宁
占地面积	3000 平方米
建筑面积	653 平方米
OPENED ON	SEPTEMBER / 2013
LOCATION	JINING / SHANDONG PROVINCE
LAND AREA	3,000 m²
FLOOR AREA	653 m²

1

2

PROTOTYPE ROOM OF SOHO TECHNOLOGY COMPANIES
SOHO 科技公司样板间

采用活力的色彩与时尚的造型，配以前卫的软装搭配，营造出信息化的时尚空间。空间划分灵活，用材追求鲜明和谐，强调材质之间肌理的对比；用色追求明快，使其整体空间呈现出数字化与科技、时尚的气势。

The design of the prototype room of technology companies adopts vivid color and fashionable modeling with avant-garde soft decoration to create an fashionable space of information technology. The design enjoys flexible space division, distinctive and harmonious material selection to emphasize contrast of texture among different materials. Bright color scheme make the entire space present digitized, technological and fashionable atmosphere.

1 济宁万达华府SOHO科技公司样板间
2 济宁万达华府SOHO科技公司样板间
3 济宁万达华府SOHO科技公司户型图

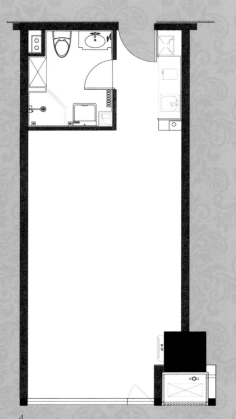
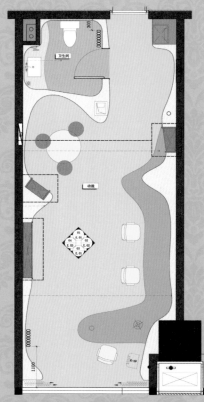

PROTOTYPE ROOM OF ANIMATION COMPANIES
动漫公司样板间

样板间的设计用线条作为主要的设计语言,追求大块面、大手笔的立面造型,运用流线型的手法,使视觉效果富有节奏韵律感,提升了空间的进深层次,回归了空间的和谐统一。采用明快的色调与时尚的软装搭配,感受现代与温馨。丰富的灯光设计,用色彩活跃空间气氛。利用线条强化流动性,营造一个具有亲和力的空间。

The design of the prototype room of Animation Company takes lines as the main design language to pursue large-scale facade modeling and adopt smooth design line to create rhythmical visual effect, enriching structure of the depth and achieving harmony and unity of the space. Color scheme with fashionable soft decoration creates modern and warm atmosphere while lighting design with rich colors strengthen sense of vitality of the space. The design uses lines to enhance smoothness, building a pleasant and attractive space.

4 济宁万达华府动漫公司样板间户型图
5 济宁万达华府动漫公司样板间
6 济宁万达华府动漫公司样板间

WENZHOU PINGYANG WANDA PALACE: PROTOTYPE ROOM
温州平阳万达华府－样板间

开放时间	2013 / 08
建设地点	浙江 / 温州
占地面积	1862 平方米
建筑面积	1328 平方米
OPENED ON	AUGUST / 2013
LOCATION	WENZHOU / ZHEJIANG PROVINCE
LAND AREA	1,862 m²
FLOOR AREA	1,328 m²

1

PROTOTYPE ROOM
样板间

此案的室内空间构筑形态采用了穿插和加减的设计手法，把靠近客厅的书房作为一个灵活的空间处理，扩大了客厅的视觉感受；餐厅北侧营造出一个红酒品赏区，与厨房相连的部分采用吧台作为半隔断，造型上高低错落，空间丰富；主卧采用玻璃隔墙，有效地扩大了视觉空间感，材料通过雅士白大理石、毛皮、玻璃、影木和镜钢等材料，体现点、线、面等概念元素，色彩明丽，与软装华丽配饰相呼应。

The structure of indoor space of this project is designed in an alternative and combined-and-separate mode. In this design, study adjacent to the living room is treated as a flexible space to enlarge the visual effect of the living room. A certain space north to the dining room is separated out as a wine bar, and the part connected with the kitchen is designed to adopt a bar counter as a semi-partition structure, creating a rich space with various modeling. Glass partition wall is adopted in the master bedroom to effectively strengthen the visual effect. The embodiment of concept elements like point, line and plane work in concert with gorgeous ornaments of soft decoration through Ariston marble, fur, glass, hardwood and stainless steel, etc.

1 温州平阳万达华府港式样板间户型图
2 温州平阳万达华府港式样板间客厅
3 温州平阳万达华府港式样板间书房
4 温州平阳万达华府港式样板间主卧

CHONGQING BA'NAN WANDA PALACE: SALES OFFICE
重庆巴南万达华府-售楼处

开放时间	2013 / 11
建设地点	重庆
占地面积	3000 平方米
建筑面积	1961 平方米
OPENED ON	NOVEMBER / 2013
LOCATION	CHONGQING
LAND AREA	3,000 m²
FLOOR AREA	1,961 m²

ARCHITECTURE OF THE SALES OFFICE
售楼处建筑

售楼处外立面结合集团标准化模板，运用现代手法，以全新的思维及手法追求"现代"美感；大面积采用镂空的锥形铝板，给人通透、立体的感觉；结合门头玻璃内的金色雕花铝板，彰显着万达特有的尊贵和豪华。整体造型折线性展开，形成城市的主要视觉交点。

Referred to standard mould of Wanda Group, facade design of the sales office adopts brand new concept and methods to pursue a "modern" sense of beauty. Hollow conical aluminum plates, which are used in large areas, create a transparent and three-dimensional effect, integrating with golden carved aluminum plates in the glass of door lintel to express the unique dignity and luxury of Wanda Group. The overall design is linearly spread out, becoming main visual focus in the city.

1 重庆巴南万达华府售楼处夜景
2 重庆巴南万达华府售楼处大堂

审图号：GS（2014）2552号

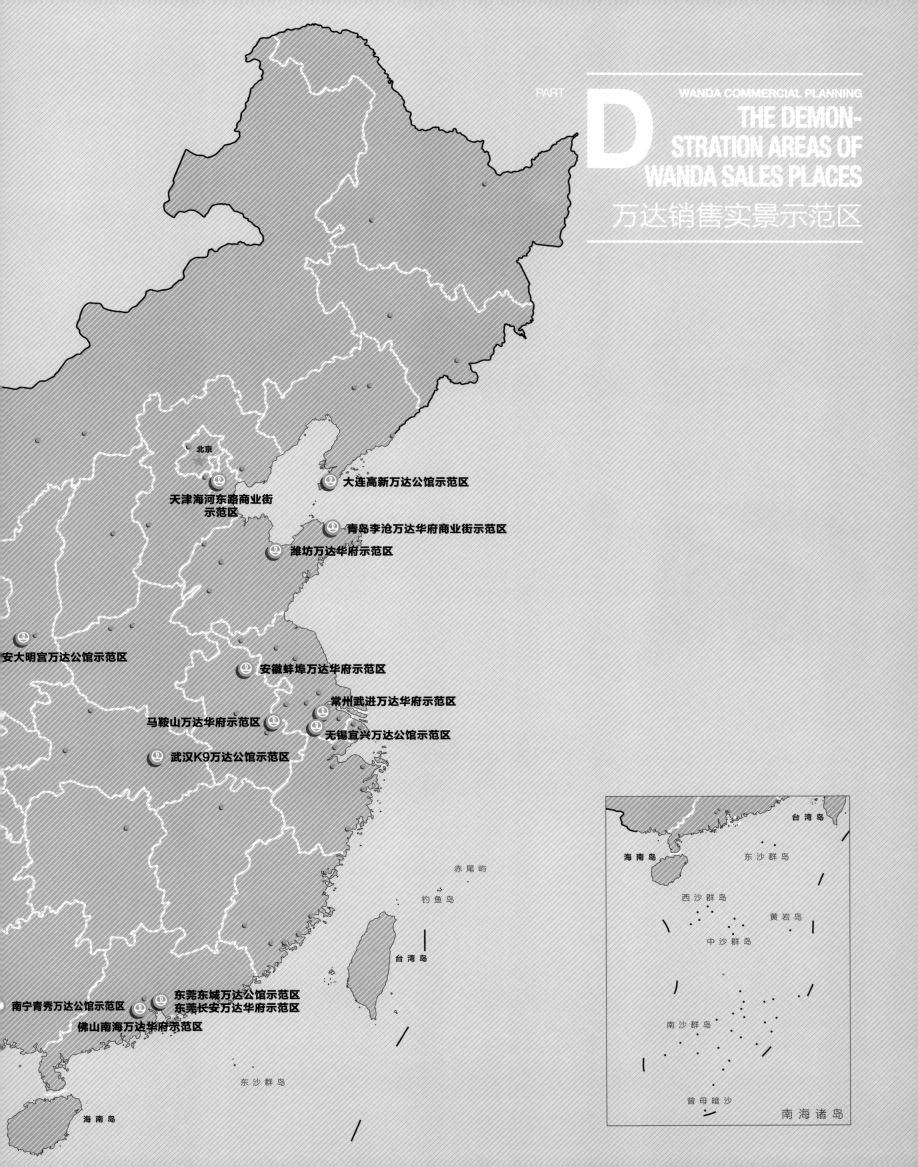

STRATEGY OF THE DEMONSTRATION AREA OF WANDA SALES PLACES

万达销售实景示范区展示策略

文／万达商业地产设计中心南区设计部副总经理 黄建好

实景示范区是实景体验的重要工具，是给客户展示项目品质效果、品牌实力及自身未来生活的体验场所。

良好的示范区展示效果，可以将客户脑海中理想化生活形态以现实版的方式展现，有效化地提升客户的购买欲望，强化客户对未来产品品质、服务质量、生活配套等居住要素的切身体验，树立客户对于品牌产品的信心，从而起到缩短客户购买决策过程及营销推广的作用（图1）。

The demonstration area is an important tool to experience the actual scenery and it is also the experience place to show clients the quality effect of project, the power of brand and their future life.

Excellent display effect in the demonstration area could help clients transfer the image of idealized life in their mind into real. It is effectively promoted clients' desire of purchasing and reinforced their personal feeling towards product quality, service and supporting living facilities in the future. Their confidence for the brand can facilitate marketing role and shorten their purchase process (Figure 1).

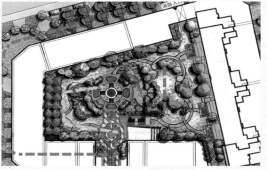

（图1）万达集团住宅销售动线典型案例

I. DESIGN OF VISIT FLOW LINE IN THE DEMONSTRATION AREA

A natural and smooth flow needs to be reasonably organized according to clients' psychological changes and demand. The flow is designed to display complete and reasonable content, preventing backtracks. Electro-mobile is required for a long flow.

一、示范区参观动线设计

根据客户心理变化及心理需求合理组织动线，让动线自然流畅。在动线设计中充分考虑展示内容的完整性、合理性，避免客户走回头路。动线过长时需考虑配置电瓶车。

售楼处→沿街底商→小区入口大门→景观展示区→住宅大堂→住宅样板间→返回售楼处（图2）。

II. FLOOR STORES ALONG STREET AND DECORATION HIGHLIGHTS

1. FACADE OF FLOOR STORES IN THE BUILDING
The facade of floor stores is in simplified European style that steady column base in dark color, elegant fust and delicate window head and cornice which constituting a simple three-phase facade. Column base uses stone in dark gray or dark brown and fust uses stone in beige. Appropriate groove with mounted

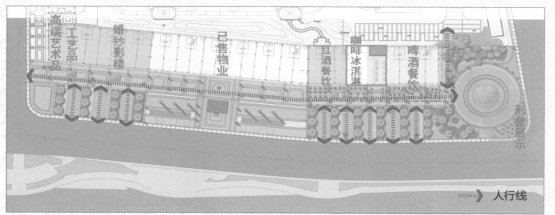

（图2）动线图

二、沿街底商及美陈展示亮点

1. 建筑底商立面

底商立面风格选取简欧风格,以深色稳重的柱座、典雅的柱身、精致的窗楣、檐口,构造出简约的三段式立面;柱座选用深灰或深棕色石材,柱身选用米黄色系石材;柱身装饰比例适度的凹槽并镶嵌仿铜云石壁灯,精湛细腻的工艺与间隔有序的节奏是形成商业街区氛围的基石;门扇上方预留不小于1米高店招位;柱式间檐口可以选用石材,也可以选用GRC造型、真石漆饰面;檐口的高度必须考虑不得遮挡上部住宅外窗的采光与视线;众多的细节是万达多年经营商业地产的精华沉淀(图3)。

(图3)底商样板段案例

2. 立面美陈

美陈包括雨篷、侧招、店招、橱窗等分类。底商立面中点缀着造型各异的牛津帆布雨篷,坡型、半球型、遮阳装饰并重的折叠型等,雅致的酒红、墨绿色调帆布,继承于欧洲百年店铺的优雅特质;不同尺度的侧招,映衬的灯箱画面,向沿街走过的人群展示着万达底商的繁华;门楣上的店招采用底板加发光字,业态与侧招呼应;窗贴与实体橱窗模拟着各式各样的业态,既是样板又是精心描摹的商业街实景(图4)。

(图4)立面美陈案例

imitated bronze marble wall lamp are applied for fust decoration. Exquisite and fine process and rhythm in orderly interval are foundation to form the atmosphere of commercial street. At least the 1m height is reserved above door leaf for signage. Cornice in column type can choose stone, GRC board or mineral varnish facing. Cornice height should not avoid the light vision of reserved window on the top of the stone. These details are accumulative experience of Wanda for many years in its operating commercial properties (Figure 3).

2. FAÇADE DECORATION

There is decoration for canopy, side signage, signage and showcase. Canvas canopy in different shapes are dotted in façade of floor stores, including slope, hemisphere and folded types with functions of sunshade and decoration, etc. Elegant canvas in wine red and dark green inherit elegance of European stores with the history of a hundred years. Side signage in different sizes and light box shows passengers the prosperity of Wanda floor stones. Shining signage above lintel and various industries in the showcase are both prototype and live scene on the commercial street (Figure 4).

III. THE GATE DISPLAY AT THE ENTRANCE OF COMMUNITY

The entrance of community is the first stop of the landscape demonstration area and also the important connection between internal and external environment of the community. Entrance is designed in classical architectural layout combined with modern function of residential community. Façade is in classical symmetric layout with three arches emphasizing central axis to separate pedestrians and vehicles. Façade column uses luxurious Corinthian order or elegant Ionic column. Luxurious and magnificent gate enhances clients' confidence for purchase, displays the power of Wanda and shows owner's distinguished identity (Figure 5).

IV. THE DISPLAY OF THE LANDSCAPE DEMONSTRATION AREA

The display of the landscape demonstration area is a way of communicating with clients in full aspects of sense. Full consideration of marketing demand

三、小区入口大门展示

小区入口是景观示范区的第一参观滞留点，同时也是小区内部与外部环境连接的重要节点。入口的设计上采用古典建筑布局结合了现代住宅小区的功能需求，立面构成为强调中轴线的三拱门古典对称式布局，达到人车分流的目的。立面柱式采用奢华的科林斯柱式或优雅的爱奥尼柱式。奢华大气的门头给客户增加购买的信心，传递了万达的实力，彰显业主的尊贵身份（图5）。

（图5）万达公馆住宅大门典型案例

四、景观示范区展示

景观示范区的展示是与客户听觉、视觉、嗅觉、味觉的多重沟通。充分考虑营销需求，有效组织客户的参观路线，让客户感受未来家园的温馨，给客户更多的想象空间，更多期望。

1. 主入口区景观展示

入口通道选用高品质石材铺装，暖色系拼花设计，简洁大气。入口通道对称式设计，两排4米高欧式多头灯柱与建筑风格相呼应，彰显尊贵品质。灯柱两侧配6~8米高的高大乔木，端头对景通常为主题雕塑或景石、水景、绿植，成为进入景观展示区的视线焦点，两侧高大乔木下搭配多层次灌木及时花形成热烈的气氛（图6）。

2. 组团内部景观展示

园区内部园路铺装可采用红色陶土砖加石材收边，自然曲线的选型设计。水系、木平台，延续高品质的同时强调温馨、生态的概念。节点设计充分考虑客户感受，运用景观手法设置视线停留点，花海、陶罐、

and effective organization of client's visit make clients feel comfort of future home and bring them more imagination and expectation.

1. THE LANDSCAPE DISPLAY AT THE MAIN ENTRANCE

The entrance channel where is simple in flower pattern design paved with high quality stone in a warm tone is in symmetric design that 4m multipoint European style lamp standards in two rows work in concert with architectural style, displaying dignity quality. 6m-8m tall trees are planted on both sides of lamp standards. The end is usually themed sculpture, stone, waterscape or green plants, which is visual focus when entering into landscape display area. Shrub in many layers and seasonal flowers are decorated under high trees to form lively atmosphere.

2. THE LANDSCAPE DISPLAY FOR BUILDING GROUPS

Road in the area is paved with red clay bricks surrounded by stone on the boundary line in a naturally curved style. Water system and wooden platform emphasize warm and ecological concepts by continuing high quality. The clients' feeling has been fully considered in allocating landscape points, including sea of flowers, clay pot, and small animals, which make the eye line of client flow with the landscape. Two sides of road are designed by micro-topography, which combines with colorful plants, multi-layers to express landscape theme. Blossom time and display effect are the consideration factors in choosing types of plants. The local growing well trees and flowers in elegant colors like pink or purple are the best choices (Figure 7).

（图6）住宅主入口案例

小动物等，让客户的视线随景观的设计流动。园路两侧采用微地形的处理手法，结合植物多色彩、多层次的设计手法表达景观主题，在植物种类的选择上应充分考虑开放时间及展示效果，骨架树种应为当地长势良好的植物；时花的颜色应选择粉色、紫色等雅致的色系（图7）。

（图7）示范区景观实景展示案例

3. 景观示范区的氛围营造

氛围包装，主要用于营销主题的表达、情景的设计及景观主题概念的提升。在样板段设计之初，策划客户参观动线的同时组织氛围包装方案。围挡画面应为自然风景画面；结合营销主题沿途布置节点内容，如儿童乐园、聚会空间、户外SPA、读书花园等，通常可以让客户参与或驻足的位置（图8）。

五、住宅入户空间展示

客户将伴随参观路线来到住宅入户空间，满足客户在较长时间室外活动的前提下，希望归于室内家园的迫切感受。住宅入口采用大理石贴面、紫铜大门，尺度大气，给客户产生尊贵感。满足客户对于未来室

3. CREATING ATMOSPHERE IN THE LANDSCAPE DEMONSTRATION AREA

Atmosphere creation is mainly used to express marketing theme, design situation and promote landscape theme concept. At the beginning of designing prototype section, atmosphere design needs to be organized with visit line. It is natural scenery picture on hoarding. Combined with marketing theme, landscape points are allocated along the way for their participation, such as children's playground, party space, outdoor SPA, reading garden, etc. which make clients stop and join in (Figure 8).

V. THE DISPLAY OF INTERNAL SPACE FOR THE HOUSE

After having outdoor activities for a long time, clients are urgent to visit indoor space. Marble facing and magnificent red copper gate at entrance bring dignity to the clients. Clients are satisfied with the requirements for safe and luxurious life in the future. Dense trees on both sides of gate create the feeling of outdoor foyer.

VI. THE RESIDENTIAL LOBBY

The residential lobby is raised double layers in the simplified European style which is consistent with lobby style and material color displayed in the temporary prototype room. The gate of entrance is a

（图8）氛围包装的伞座软装及草坪白鸽案例

double purple bronze gate by 3.6m high and 3m wide. Façade of the lobby is the key point which is in the European style with squared stone column, painted

(图9)住宅大堂案例

内居住空间安全、奢华、唯美的生活需要及心理需要。在门头两侧要利用茂密的种植层次形成绿色围合空间,形成室外玄关的空间感受。

六、住宅大堂

实景住宅大堂为双层挑高,风格采用简欧装饰风格,同临时样板间展示时的大堂风格及材料色调保持一致;住宅入口大门采用3.6米高、3米宽的双开紫铜色大门;大堂正立面为重点设计墙面,用石材装饰的方柱结合大幅绘画的主背景墙面,造型天花,两层叠级、华丽的水晶吊灯;整体的设计手法为简欧风格,根据成本造价要求,立面部分局部采用石材线条,大面积使用双色仿石材面砖,衔接位置用香槟金色的线条收口(图9、图10)。

background wall, shaped ceiling and magnificent two-layer crystal chandelier. To match the costly requirement, partial façade uses stone and two-tone face bricks in large area and the connection position is filled with golden line (Figure 9 and Figure 10).

The important significance of live demonstration area is to promote sales as a model, which can bring real feeling to clients about the living environment in the future. The display can better improves clients' recognition of the residence, and better demonstrate the supreme quality of Wanda.

(图10)住宅大堂案例

销售实景示范区的重要意义就是促进销售,起到样板的作用,让客户对未来的居住生活环境有更加真实的感受,通过展示提升客户认知,更好地展示万达品质。

DEMONSTRATION AREA OF DALIAN HIGH-TECH ZONE WANDA MANSION

大连高新万达公馆示范区

开放时间	2013 / 08
建设地点	辽宁 / 大连
占地面积	10118 平方米
建筑面积	6209 平方米
OPENED ON	AUGUST / 2013
LOCATION	DALIAN / LIAONING PROVINCE
LAND AREA	10,118 m²
FLOOR AREA	6,209 m²

PROJECT OVERVIEW
项目概述

大连高新万达公馆位于高新区七贤东路南端沿海地块，是大连目前唯一一拥有绝佳山海资源同时又享有顶级城市商业配套的项目。

Dalian High-tech Zone Wanda Mansion is located at a littoral parcel of the south end of Qixian East Road, High-tech District of Dalian. It is the only project which contains superexcellent mountain, ocean resources and top grade urban commercial facilities in Dalian.

1

2

1 大连高新万达公馆动线图
2 大连高新万达公馆入口
3 大连高新万达公馆会所建筑外立面

ARCHITECTURAL PLANNING
建筑规划

设计采用19世纪欧洲新古典主义，以"古典美学"为理论支撑，强调空间造型的等级感、序列感与仪式感，强化轴线对景，讲求构图的形式美感，多为对称式，秩序严谨，体现出居住的身份感与居所的价值感。

This project is designed in European neoclassicism style of the 19th century which is based on the theory of "classical aesthetics" to emphasize the senses of grade, sequence and ceremony of space modeling, reinforce symmetric landscape of axis and focus on form aesthetic, mostly symmetric style, and rigorous sequence so that to reflect senses of residence status and residence value.

4　大连高新万达公馆11#楼建筑外立面
5　大连高新万达公馆11#楼建筑外立面
6　大连高新万达公馆11#楼细部
7　大连高新万达公馆会所宝瓶柱
8　大连高新万达公馆会所宝瓶柱
9　大连高新万达公馆别墅剖面图

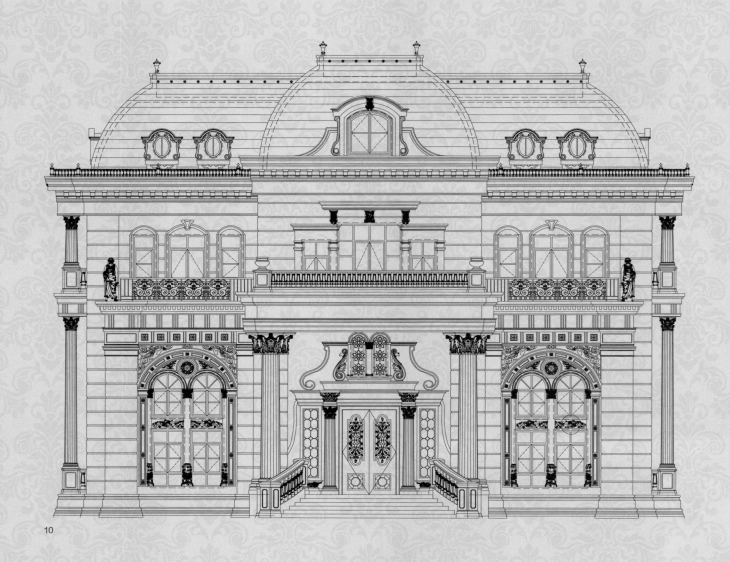

10

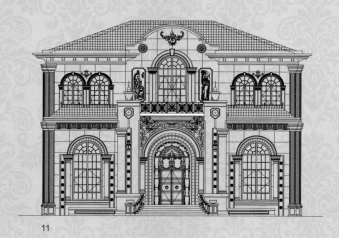

11

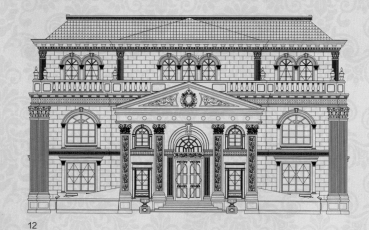

12

外形丰富而独特，形体厚重，贵族气息在建筑的冷静克制中优雅地散发出来。建筑线条鲜明，凹凸有致，尤其是外观造型独特，呈现出一种华贵气质。法式建筑的另一个特点，善于在细节雕琢上下工夫，呈现出浪漫典雅风格。法式建筑的三段式结构气势恢宏，于威严之中透露高贵与活力，备受追求品质的上层人士所喜爱。

Rich appearance, dignified volume and noble odor are aroused gently from the clam and restrained architecture. Vivid lines, distinct texture and especially unique appearance present sumptuous and magnificent temperament. One feature of the French architecture is extraordinary detailed carving, which presents romantic and elegant style. Three-order structure of the French architecture looks grand and grandiose with dignity and dynamic, which is elites' favorite.

13

14

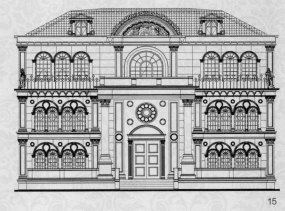
15

16

10 法国古典风格别墅立面图
11 西班牙风格别墅立面图
12 佛罗伦萨风格别墅立面图
13 巴洛克风格别墅立面图
14 枫丹白露风格别墅立面图
15 威尼斯风格别墅立面图
16 英国浪漫风格别墅立面图

17

18

19

PROTOTYPE ROOM OF THE FINISHED MANSION
实楼样板间

一层为宴会殿堂，是海屿墅的入户层，均使用大开间、大进深，打造出大气十足的空间感，形成超大观海客厅空间。客厅到餐厅自然连接，形成互不干扰的完美动线。二层为家庭私属空间，户型方正，主卧私密性强。双套房及宽敞家庭厅，为家庭成员准备了丰富舒适的空间。三层为主人海景套房，专为主人量身打造，空间使用充分，房间用途设置完备；主卧室配合超大观景露台及海景浴室，全时观赏海上风光。在材料运用方面分别运用了金沙米黄、意大利黑金花、金镶玉、虎眼石、金丝白玉、贝壳马赛克、波斯玉、木纹玉石、英伦玉和古纹玉等名贵国外进口石材，配以黑胡桃等优良材质的木作。

The ground floor is a banquet hall and the entrance of sea island mansion, which adopts large opening and deep depth to create a super large living room of seascape viewing with a spacious sense. The living room naturally connects with the dining room to form a perfect passageway with non-interfering function. The second floor is the private space of family with foursquare house type. Main bedroom stresses privacy. Double suites and capacious family room provide rich and comfortable space for family members. The third floor is a master ocean-view suite, which is especially designed for owner with reasonable and advisable layout and perfect equipment. And with super large viewing terrace and ocean-view bathroom, owner can enjoy the seascape in anytime. In aspect of material application, the mansion adopts fine wood carving, such as black walnut, and precious stones from abroad, such as Italian Perlato Svevo, Italian Portoro Gold, phyllostachys jade marble, tiger's eye marble, golden cloud marble, shell mosaic tiles, wave weasley silk, Serpeggianto, Turkey stone and Ancient Vein Onyx.

20

LANDSCAPE
景观

景观设计注重细节和造型，给人以华美、精致、富丽堂皇的感觉，并运用现代的材质及工艺演绎传统文化的精髓，使景观不仅彰显尊贵和舒适，还拥有时代的特征。

The landscape concentrates on details and modeling to create senses of gorgeousness, exquisiteness and magnificence. Modern materials and deducing process of essence of traditional culture make the landscape not only represent dignity and comfort, but also possess characteristics of the time.

21

22

17　大连高新万达公馆别墅样板间客厅
18　大连高新万达公馆别墅样板间主卧
19　大连高新万达公馆别墅样板间书房
20　大连高新万达公馆别墅样板间户型图
21　大连高新万达公馆别墅样板间庭院
22　大连高新万达公馆别墅景观连廊

DEMONSTRATION AREA OF DONGGUAN DONGCHENG WANDA MANSION

东莞东城万达公馆示范区

开放时间	2013 / 07
建设地点	广东 / 东莞
占地面积	4860 平方米
建筑面积	855 平方米
OPENED ON	JULY / 2013
LOCATION	DONGGUAN / GUANGDONG PROVINCE
LAND AREA	4,860 m²
FLOOR AREA	855 m²

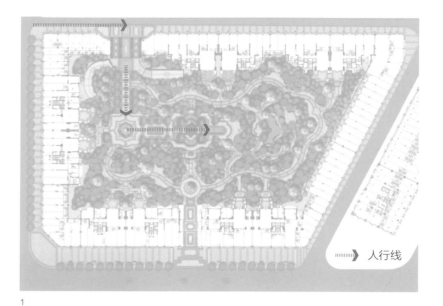

人行线

1

ARCHITECTURAL PLANNING
建筑规划

项目主要采用Art-Deco的新古典主义风格立面，充分展现了建筑古典美和现代感的交融。外观采用三段式设计，层次分明，不拘一格，彰显出建筑的个性与不可复制性，打造为东莞首席城市居所。小区大门重点体现大气奢华的公馆品质，东区大门为三拱式，两侧为独立车行出入口，中间为人行出入口，做到人车分流。立面造型注重原始的古典尺度比例，尊重历史元素；细节地刻画出高贵、华丽的入口形象。

The project mainly adopts neoclassical facade of the Art-Deco style to fully display the melting of classical beauty and modern sense. The facade adopts three-order design with distinct layers and various patterns to present its personality and uniqueness for making the mansion become the first urban residence in Dongguan City. The gate of entrance of the community concentrates on reflecting grandiose and luxurious quality of the mansion. The east gate is designed in three-arch style with independent vehicle entrances on both sides and pedestrian entrance in the middle to separate pedestrians and vehicles. The facade modeling focuses on the original classical scale ratio which respects the historical elements to delicately depict the noble and superb entrance image.

1 东莞东城万达公馆动线图
2 东莞东城万达公馆西大门
3 东莞东城万达公馆海神主雕塑
4 东莞东城万达公馆天使雕塑
5 东莞东城万达公馆儿童雕塑

6

商业裙楼立面在体现住宅建筑的群楼效果的同时充分考虑与周围环境相协调，采用了追求华贵、高雅的新古典立面建筑风格。商业立面主要以高低错落、疏密有致的圆拱形铺面组合而成；整体屋面均采用坡屋顶，以此增强整个裙楼商业的完整性。立面上，一方面保留古典欧式建筑中的材质、色彩的大致风格，可以很强烈地感受传统的历史痕迹与浑厚的文化底蕴，就像欧洲的传统贵族一样，讲究庄重、华贵；另一方面，又摒弃了过于复杂的肌理和装饰，简化了线条，通过现代幕墙、金属构件和色彩艳丽的金属雨篷，来迎合现代人的审美情趣。

The facade of commercial podium building adopts neoclassical architectural style, which pursues luxury and elegance, with the consideration of reflecting the effect of building groups of residential housing and coordinating with surrounding environment at the same time. The roof is designed in pitched roof style to enhance the integrity of the whole podium building, and the facade is designed in series of arched surfacing in staggered arrangement and appropriate density. On one hand, the facade keeps the general style of texture and color in the European classical architecture to make residents feel historical rudiment of traditional architecture and rich cultural deposits, and just like European noble people who pursue solemnity and luxury. On the other hand, the facade abandons excessively complicated texture and decoration and adopts simplified line style with modern curtain wall, metal construction and flamboyant metal awnings to cater the aesthetic taste of modern people.

6 东莞东城万达公馆底商建筑外立面局部
7 东莞东城万达公馆西大门立面图
8 东莞东城万达公馆底商建筑外立面

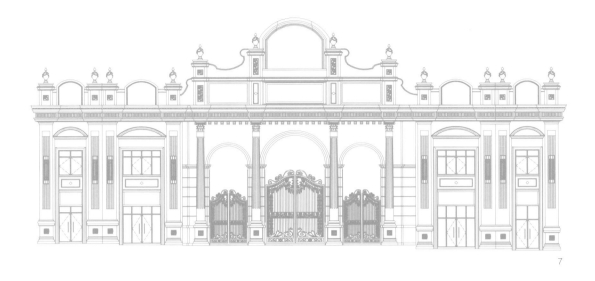

7

8

PROTOTYPE ROOM OF THE FINISHED MANSION
实楼样板间

以万达公馆欧式新古典装修风格为蓝本，并进行升级，更倾向于居家实用——稀缺的环境资源、至臻的建筑品质、近乎完美的户型设计、殿堂级纯大户奢阔空间、总统级双主卧套房配置、宫殿级超大客厅、欧洲皇室专享尊贵古铜门、罕有的精品酒店式家居管理……这一切，都给人温暖、浪漫及永恒的感觉。

Based on the European neoclassical style of Wanda Mansion, the prototype room is designed in a more practical way for living at home. All of these make residents feel warm, romantic and eternal, including scarce environmental resource, excellent architecture quality, nearly perfect housing type design, palatial and capacious space, president suite with double master rooms, palatial and super-large living room, honorable copper door of the European royal style, home management of rare boutique hotel type, etc.

9 东莞东城万达公馆景观雕塑
10 东莞东城万达公馆样板间入户大堂
11 东莞东城万达公馆样板间户型图

12

13

14

12 东莞东城万达公馆样板间客厅
13 东莞东城万达公馆样板间主卧
14 东莞东城万达公馆样板间次卧
15 东莞东城万达公馆甲级写字楼样板间
16 东莞东城万达公馆样板间户型图

15

PROTOTYPE ROOM
OF FINISHED 5A OFFICE BUILDING
实楼甲级写字楼样板间

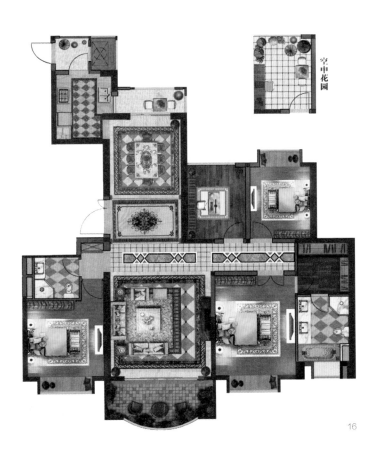

空中花园

16

时尚、简约、现代感,是该项目的设计主导。现代感的建筑神韵在内装上得以延伸,搭配富有文化气息的工艺品与装饰画,使整体空间低调而不失奢华;合理的空间功能布局与划分,充分满足了人性化的需求;丰富细腻的空间细节处理、典雅而不失亲切的材质感官体验、专业精致的灯光氛围设计,共同准确地渲染出时尚感与现代感相结合的艺术氛围。

Fashionable, simple and modern senses are the leading idea of the project design. The modern sense is reflected in the interior design, along with cultural artwork and decorative pictures by making the entire space humble without losing luxury. Reasonable space layout and advisable space partition fully satisfy humanized demand. Rich and delicate space detail treatment, texture sensory experience of elegance without losing geniality and professional and exquisite lighting atmosphere designing precisely, all of these render the art atmosphere of combination of senses of fashion and modernity.

DEMONSTRATION AREA OF TIANJIN HAIHEDONGLU COMMERCIAL STREET
天津海河东路商业街示范区

开放时间	2013 / 08
建设地点	天津
占地面积	13780 平方米
建筑面积	22252.63 平方米
OPENED ON	AUGUST / 2013
LOCATION	TIANJIN
LAND AREA	13,780 m²
FLOOR AREA	22,252.63 m²

1 天津海河东路商业街建筑立面夜景
2 天津海河东路商业街动线图
3 天津海河东路商业街前广场欧式车马雕塑

ARCHITECTURAL PLANNING
建筑规划

立面法式新古典风格承续了天津开埠通商、欧风东渐的地域历史文脉。设计采用经典的欧式建筑三段式的基本格式，将法式孟莎盔顶、柱式、拱券、线条等古典元素以恰当的布局和比例组合在一起，凸显欧式建筑的气势恢宏，庄严大气之态。

The facade of the French neoclassical style inherits the historical context of Tianjin City in the period of opening commercial intercourse and emergence of the European style. The design adopts the classic three-order style of the European architecture with the appropriate layout and proportion of the classical elements such as the mansard roof, columns, arches and lines to highlight the magnificence and solemnity of the European style architecture.

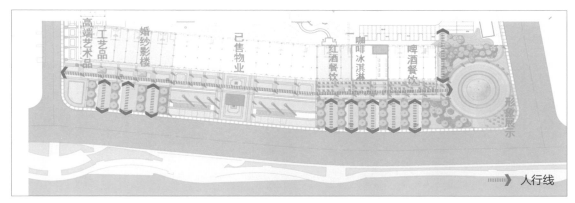

天津海河东路商业街滨河整体夜景

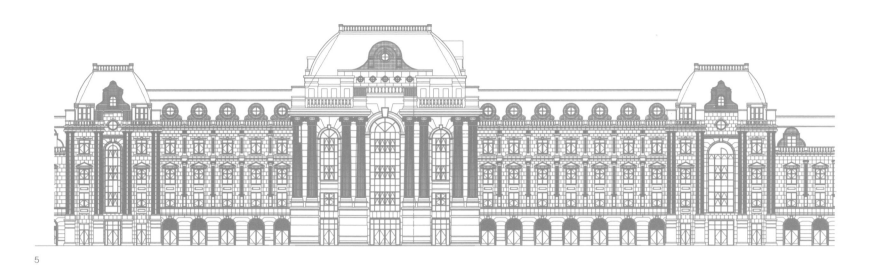

5

5 天津海河东路商业街立面图
6 天津海河东路商业街夜景
7 天津海河东路商业街前广场建筑立面
8 天津海河东路商业街前广场人物雕塑
9 天津海河东路商业街前广场人物雕塑

6

LANDSCAPE OF COMMERCIAL STREET
商业街景观

场地设计充分考虑人流动线规律，端部设街角广场吸纳人流，中部设主广场形成开放活动空间聚拢人流。结合广场分别设置喷水池、标志柱等景观小品，凸显场地的个性，增强场所感，丰富场地空间。整个景观场地采用中轴对称布局，景观小品均为欧式造型，与建筑布局和风格配合辉映，相得益彰，呈现出法度严谨、气势恢宏的欧式园林特色。

For convening people, the design takes full consideration of pedestrian and vehicle behaviors to set up a main square of opening activity space and corner squares. According to the square design, garden ornaments, like fountains and columns, set up to highlight the plot characters, reinforce the sense of commercial atmosphere and enrich the space design. The landscape design adopts axial symmetry layout, and garden ornaments are designed in European style, bringing out the best in each other to appear elaborate and magnificent European garden features.

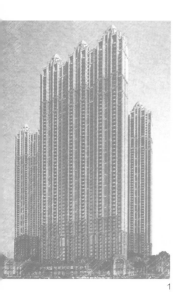

DEMONSTRATION AREA OF NANNING QINGXIU WANDA MANSION
南宁青秀万达公馆示范区

开放时间	2013 / 12
建设地点	广西 / 南宁
占地面积	14478 平方米
建筑面积	16430 平方米
OPENED ON	DECEMBER / 2013
LOCATION	NANNING / GUANGXI ZHUANG AUTONOMOUS REGION
LAND AREA	14,478 m²
FLOOR AREA	16,430 m²

ARCHITECTURAL PLANNING
建筑规划

南宁青秀万达公馆注重本身的连续性，建筑形象追求完整协调，营造出具有整体文化特色和历史文脉的城市空间环境，以优美的形象屹立于城市之林。立面的设计原则是特点鲜明、协调统一，采用Art-Deco建筑风格，强调竖向线条，结合古典三段式的经典构图手法，再配以Art-Deco经典元素，形成具有鲜明特色和标志性的建筑形象。

Nanning Qingxiu Wanda Mansion concentrates on succession of the buildings and pursues integrity and coordination of architectural image to create urban space environment of cultural features and historical context, standing among cities with graceful image. The design principle of facade is distinct features and uniform coordination. Facade adopts the Art-Deco style which focuses on vertical lines, incorporating with classical structural method of three orders style and classic elements of the Art-Deco style to build a landmark with distinctive features in this area.

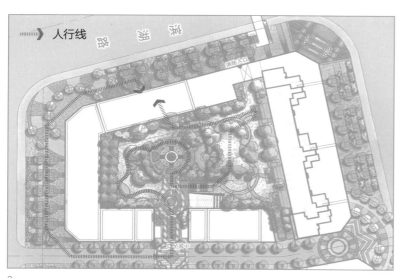

LANDSCAPE
景观

项目的景观设计，意在彰显皇家贵族的体验。住宅、商业的景观细节都同建筑外饰风格统一。中轴线上的骑士雕塑、中庭亭子和跌水景都体现出意式皇家的风范，十根模数制的罗马柱以及柱上的石狮子，精巧地雕琢出贵族品质。外围商业街配合整个建筑的意式风情，通过不同面层处理、搭配异国情调的遮雨棚与花钵小品，彰显出高贵典雅的贵族气息。

The landscape design is meant to enhance royal and aristocratic experience. The details of landscape of residential and commercial areas share the same style with architectural exterior decoration. Knight sculpture, atrium pavilion and cascade water feature on the central axis perfectly interpret Italian royal style, and ten proper Roman columns with guardian lions elaborately depict noble quality. The peripheral commercial street with the Italian-style architecture exudes noble and elegant aristocratic scent through various architectural surface processes with exotic awnings and feature pots.

1 南宁青秀万达公馆建筑外立面
2 南宁青秀万达公馆示范区动线图
3 南宁青秀万达公馆示范区水景
4 南宁青秀万达公馆示范区景观
5 南宁青秀万达公馆示范区中庭景观

6

7

8

6 南宁青秀万达公馆示范区景观台阶
7 南宁青秀万达公馆示范区景观雕塑
8 南宁青秀万达公馆商业街效果图
9 南宁青秀万达公馆商业街景观
10 南宁青秀万达公馆商业街夜景

COMMERCIAL STREET
商业街

住宅底商采用简洁的线条结合局部坡屋顶收分的手法。项目整体感强烈，轮廓清晰，注重文化和艺术内涵的表达，以及对品质感的极致追求。住宅大门的设计融入到整个设计当中，并且予以强化，增加立面装饰，配以古典景观小品和铺地形式，形成重要的景观节点。

The floor stores of residential area adopt simplified lines with partial pitched roofs in different dimensions. The whole structure possesses intensive integrity and distinct sketch, focusing on the expression of culture and artistic connotation and the ultimate pursuit of sense of the quality. The design of entrance gate blends in the whole design work with enhancive facade decoration integrating with classical garden ornaments and paving form to form important landscape nodes.

DEMONSTRATION AREA OF WUHAN K9 WANDA MANSION
武汉 K9 万达公馆示范区

开放时间	2013 / 07
建设地点	湖北 / 武汉
占地面积	44140 平方米
建筑面积	地上 311900 平方米
	地下 81980 平方米

OPENED ON	JULY / 2013
LOCATION	WUHAN / HUBEI PROVINCE
LAND AREA	44,131 m²
FLOOR AREA	311,900 m² ABOVE GROUND
	81, 980 m² UNDERGROUND

ARCHITECTURAL PLANNING
建筑规划

武汉K9万达公馆位于武汉中央文化旅游区的北侧，建筑为Art-Deco风格，注重功能又不失装饰性。线条形式在竖向灵活运用，通过重复、对称、渐变，凸显了立面的高大效果。立面采用对称性和"三段式"的构图原则，顶部层层收缩呈阶梯状，强调垂直线条，形成类金字塔的型体，利用整体凹凸表现高耸挺拔之感。立面材质以浅色为主，多采用米黄、咖啡色的天然石材，与玻璃幕墙相结合。

Wuhan K9 Wanda Mansion, which is located at the north of central cultural tourism area in Wuhan City, was designed in the Art-Deco style with concentrating on both function and decoration. Lining form of vertically flexible application presents towering effect of facade through the techniques of repetition, symmetry and gradient. The facade adopts symmetry and the three-order design style with adder-shaped top to enhance vertical lines and form pyramid shape, reflecting a tall and straight sense. The facade texture is designed in light color tone with mainly adopting combination with glass curtain wall and natural stones of beige and light brown colors.

1

2

1 武汉K9万达公馆鸟瞰图
2 武汉K9万达公馆动线图
3 武汉K9万达公馆建筑外立面

LANDSCAPE
景观

武汉K9万达公馆的园林设计中将地形下沉后再抬高，营造一个有落差的跌水空间，形成一个富有变化又开放性的展示空间，结合代表欧式经典元素的水钵、花钵、雕塑，形成整体大气的入口广场氛围。延续轴线景观，水景末端点缀欧式雕塑作为入口处的对景，形成紧凑而又丰富的序列空间。中心水景区作为景观轴线的高潮部分，开阔的水景空间、雕塑、绿化和欧式构筑物有机地融合在一起。

After sinking and then elevating the landform, the landscape design of Wuhan K9 Wanda Mansion creates cascade space of various layers to form changeable and opening display room, combining with representative European classic elements such as water feature, feature pot and sculpture, to build an integrate and imposing entrance square. Along the axis landscape, the European sculptures are embellished at the end of waterscape as opposite scenery of the entrance, and make sequence space intensive and ample. As the climax of axis landscape, central waterscape area becomes expansive space, through the sculptures, greening and European structures coalesce together.

4 武汉K9万达公馆示范区中轴景观
5 武汉K9万达公馆示范区景观草坪

6

7

6 武汉K9万达公馆样板间主卧
7 武汉K9万达公馆样板间书房
8 武汉K9万达公馆样板间客厅
9 武汉K9万达公馆样板间客厅
10 武汉K9万达公馆样板间主卧
11 武汉K9万达公馆样板间衣帽间

8

PROTOTYPE ROOM OF FINISHED MANSION
实楼样板间

设计采用法式装饰风格元素，结合细腻精致的工艺，突出古典优雅的气质，运用现代的手法、材质和工艺还原古典气质和现代精神的双重审美效果；运用清淡的色彩、精美复古的造型，达到雍容华贵、细腻雅致的清新自然装饰效果。样板间客厅地面为石材拼花铺装；墙面为白色混油配搭高档花鸟主题的壁纸；天花运用线条繁复细致的装饰线组合设计元素，配以华丽水晶吊灯，体现法式贵族特有的高雅品质。

The design adopts many French decorative elements integrating with fine and exquisite processes to promote classical and elegant temperament. The modern design process and texture restitute dual aesthetics of classical temperament and modern spirit, and light color and vintage modeling achieve fresh and natural decorative effect without losing dignity and grace. The living room of the prototype room is decorated with paving parquet stones, white resin painting wallpaper of flowers and birds theme, portfolio-designed ceiling with complicate and delicate lines and gorgeous crystal chandelier to reflect peculiar and elegant quality of French nobility.

DEMONSTRATION AREA OF WUXI YIXING WANDA MANSION

无锡宜兴万达公馆示范区

开放时间	2013 / 05
建设地点	江苏 / 无锡
占地面积	12.05万平方米
建筑面积	16.33万平方米
OPENED ON	MAY / 2013
LOCATION	WUXI / JIANGSU PROVINCE
LAND AREA	120,500 m²
FLOOR AREA	163,300 m²

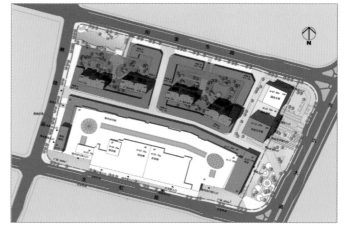

2

PROJECT OVERVIEW
项目概述

工程用地面积12.05公顷,公馆区总建筑面积16.33万平方米。其中地下室2.79万平方米;地上部分住宅10.71万平方米,底商2.78万平方米,物业及配套用房510平方米;共4栋楼,均为高层高级住宅楼。

The project land area is 12.05 hectares, and overall floor area of the mansion compound is 163,300 square meters which includes the basement, 27,900 square meters, the residential part above ground, 107,100 square meters, the floor traders, 27,800 square meters, and the property management & ancillary space area, 510 square meters. There are four buildings in total, and all of them are high-rise and high-grade residential buildings.

1 宜兴万达公馆主入口大门
2 宜兴万达公馆总平面图

3 宜兴万达公馆景观
4 宜兴万达公馆喷泉

3

4

ARCHITECTURAL PLANNING
建筑规划

本项目建筑风格吸取了类似"欧陆风格"的一些元素处理手法,但以简化或局部使用,配以大面积墙及玻璃或简单线脚构架,在色彩上以大面积浅色为主,装饰品相对简化,追求一种轻松、清新、典雅的气氛。

The project absorbs some features from "euro-continental style" and uses them in a simplified way in some local spaces in coordinating with large-area wall, glass or simple architrave framework. It adopts light color in large area with simplified ornaments to pursue relaxation, refreshment and elegance.

LANDSCAPE
景观

设计师在对凡尔赛宫细致研究的基础上,将经典进行解构和重塑,让居者在高贵、精致的景观环境之中享受古典主义园林的尊贵与静谧。在设计中运用了丰富的手法,如轴线对称、竖向变化、高差错落、疏密对比。

Basing on detailed study on Versailles Palace Garden, the designer builds the landscape by deconstructing and remodeling the classic to make residents enjoy nobleness and quietness of the neoclassical garden in this noble and exquisite landscape environment. Plentiful techniques have been used in this design such as axial structure, symmetrical layout, vertical changes, orderly and high-low arrangement, density comparison, axis openness and greening display.

5 宜兴万达公馆入户大堂
6 宜兴万达公馆客厅
7 宜兴万达公馆主卧

INTERIOR DESIGN
内装

本案汲取欧式风格精粹,体现出恢宏的气势,构建豪华舒适的居住空间,呈现出中西交融的华贵与浪漫。注重细节处理,大量运用雕花、线条描金,同时辅以雕刻精细的家具,崇尚"冲突"之美。空间秉持典型的法式风格搭配原则,深色木饰面表面略带雕花,配合顶面造型和欧式雕花的弧形曲度,显得优雅华贵。

The project absorbs essences of European style to reflect magnificent and imposing momentum, structures luxurious and comfortable living space to present resplendence and romance of the combination of the East and the West. It focuses on details by adopting carve patterns, gilded lines and exquisitely-engraved furniture to show the gorgeousness of "conflict". Adhering to the principles of typical French style, the designer uses dark wood facing with embellished carve patterns in coordinating with top surface modeling and European style carve patterns to appear elegance and sumptuousness.

DEMONSTRATION AREA OF XI'AN DAMINGGONG WANDA MANSION

西安大明宫万达公馆示范区

开放时间	2012 / 04
建设地点	陕西 / 西安
占地面积	7118 平方米
建筑面积	2321 平方米
OPENED ON	APRIL / 2012
LOCATION	XI'AN / SHAANXI PROVINCE
LAND AREA	7,118 m²
FLOOR AREA	2,321 m²

1 西安大明宫万达公馆示范区动线图
2 西安大明宫万达公馆示范区景观
3 西安大明宫万达公馆示范区中央水景设计图
4 西安大明宫万达公馆示范区水钵

LANDSCAPE
景观

西安大明宫万达公馆位于历史悠久的西安未央区，拥有浓郁的宫廷文化背景与地域特色。通过大气、对称的中轴及细致的雕刻，再现了卢浮宫古典欧陆皇家园林的景观魅力，提倡休闲自在的生活。明丽华美的景观色彩和精致细腻的景观小品，营造轻快愉悦的生活氛围；浓密绿地区域，结合欧式廊架、花钵小品等的布置，透露出浓郁的法式浪漫风情。

The mansion is located in the time-honored Weiyang District, Xi'an City, and possesses full-bodied cultural background of palace and regional characteristics. It reproduces the charm and beauty of the classical eurocontinental royal garden of Louvre to advocate leisurely and comfortable life style. Bright and ornate landscape color, exquisite and delicate garden ornaments create relaxed and delighted living atmosphere. Dense greenbelt area, integrating with the European landscape corridor, feature pots, and etc., reveals a full romantic French style.

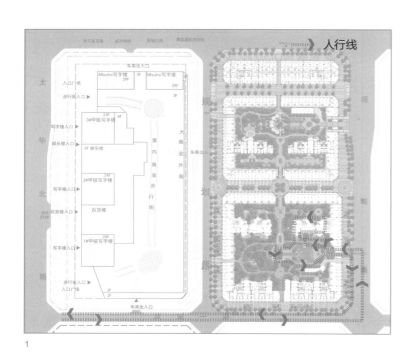

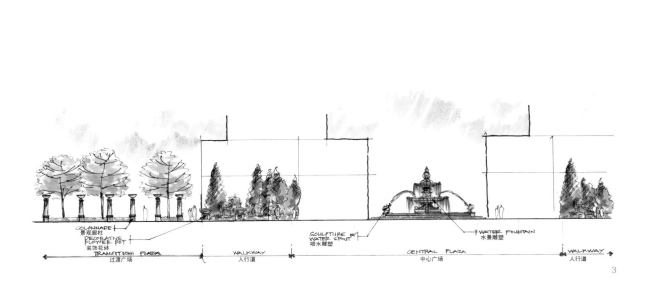

5

PROTOTYPE ROOM OF FINISHED MANSION
实楼样板间

客厅和卧室既延续法式清新、亮丽、简约的基调，形成轻盈、活泼的室内气氛，又以其对称的轴线、鲜明的线条、跳跃对比的色彩，渗透法式风格的雅致与闲适。跳跃式的软装做点缀的搭配与衬托，有力地体现其本色的基调。

The living room and the bedroom continue to adopt freshness, glow and simplicity of the French style as the key tone to create lighthearted and vivacious indoor atmosphere and permeate elegance and leisure of the French style by symmetrical axis, distinct lines and contrastive colors. Besides, embellishment of multi-style soft decoration perfectly serves as a foil to the key tone.

6

7

8

5 西安大明宫万达公馆样板间入户大堂
6 西安大明宫万达公馆样板间户型图
7 西安大明宫万达公馆样板间客厅
8 西安大明宫万达公馆样板间主卧

DEMONSTRATION AREA OF MA'ANSHAN WANDA PALACE
马鞍山万达华府示范区

开放时间	2013 / 10
建设地点	安徽 / 马鞍山
占地面积	2.1 万平方米
建筑面积	9.96 万平方米
OPENED ON	OCTOBER / 2013
LOCATION	MA'ANSHAN / ANHUI PROVINCE
LAND AREA	21,000 m²
FLOOR AREA	99,600 m²

PROJECT OVERVIEW
项目概述

马鞍山万达销售物业包含：住宅、商铺、写字楼和SOHO。住宅部分以新城市主义为规划理论，坚持公共利益优先原则，与城市周边环境协调，注重小区的归属感，优先考虑公共开放空间，维护生态环境，将小区、商业、公园整合成为一个生态、宜居、精致、人文的新型社区。

Ma'anshan Wanda Properties for Sale include residence, store, office building and SOHO. The residence property, under the planning theory of new urbanism and priority principle of public interest, emphasizes sense of affiliation, priority of public open space and maintenance of ecological environment in coordinating with urban surrounding environment to make housing estate, commercial estate and garden be merged into an ecological, livable, delicate and humanistic community.

1

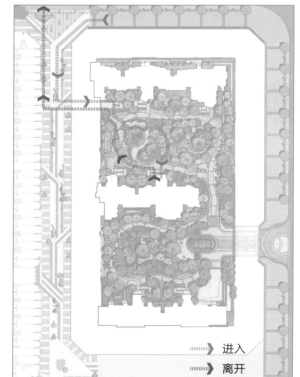

2

1 马鞍山万达华府住宅单元入口大门
2 马鞍山万达华府动线图
3 马鞍山万达华府商业街夜景
4 马鞍山万达华府商业街外立面

COMMERCIAL STREET
商业街

外墙材料采用涂料、石材、铝板等，配以凹凸窗与阳台，部分阳台采用玻璃，色彩素雅而温馨。底部墙裙采用砖石饰面，在窗及雨蓬等细节处理上采用相对现代的设计手法，使建筑彰显出一定的现代气息。以独特涵养与周围城市建筑形式相协调，既阴柔又阳刚的气息，力求创造具有时代特色的高品位建筑风格。

The exterior wall adopts various materials, such as exterior wall covering, stone and aluminum plate, integrating with built-in window, balcony and glass balcony to create simple but elegant, warm and sweet atmosphere. Bottom wainscot adopts masonry facing while window and awning adopt modern style in detail treatment, endowing the building with certain modern quality. The unique and distinct commercial street, appearing to be delicate and majestic, is in perfect harmony with surrounding urban architectural style to create high-grade architectural style with characteristics of the times.

5

LANDSCAPE
景观

景观采用欧式新古典风格，社区入口为展示形象空间，偏重礼仪形象。以中心轴线对称布局、林荫特色树以及序列性花钵小品做引导，满足人流的便达性，体现社区的欧式尊贵品质。园区内院为中心活动景观空间，其功能使用作为设计主要遵循条件，为业主打造出以欧式田园为基调的、自然、阳光、休闲的放松生活。化直为曲的漫步道贯穿整个园区，阳光交流绿地是其核心景观，精致的小品与活动小广场体现着欧式经典元素，丰富了园内人群视觉，成为园区的亮点。

The landscape is designed in the European neoclassical style. As an image of displaying space, the community entrance emphasizes etiquette and image. The symmetrical layout of central axis, tree-lined pathway and orderly arrangement of feature pots ensure convenient accessibility of visitor flow, and presents the European style dignity of the community. The central part of the garden is planned as activity landscape space, which concentrates on function and utilization. It will provide natural, sunny, casual and relaxed European idyllic living environment for the residents. The promenade winds through the whole garden. The communication of sunshine and green space is the core landscape. As the highlight of the garden, delicate ornaments and small event square, which are classic elements of the European garden, provide rich visual enjoyment for residents.

5 马鞍山万达华府实景示范区鸟瞰
6 马鞍山万达华府实景示范区
7 马鞍山万达华府实景示范区

PROTOTYPE ROOM OF FINISHED PALACE
实楼样板间

采用清新的新古典设计风格和质感丰富的材质，凸显室内细腻、典雅的气质，让空间得以灵活地转换，使人既新奇又舒适。宽敞的客厅借清爽的壁纸与质地温润的木饰面，使墙面既有节奏感又有协调感。为丰富餐厅空间的延展性，采用了装饰性强的蚀花玻璃与优雅的装饰壁灯，共同营造了适宜的就餐环境。主卧室为提升整体品质感，选择了整体宽幅的纯手绘壁纸，细节处理上运用了欧式柱、雕花和线条，工艺精细考究；色彩处理上，采用了大面积的暖色与冷色搭配的组合。

The simple and elegant neoclassical design style and textured materials present exquisite and elegant quality of interior environment of the prototype room, that create flexible space and make people feel fresh and comfortable. Refreshing wallpaper and mild timber facing make walls of the capacious living room more rhythmic and harmonious. For enriching extensibility and ductility of the space, decorative pattern glass is adopted in the dining room, along with elegant decorative wall lamp to create comfortable dining environment. Meanwhile, for promoting a sense of quality, the master room chooses broad and hand-painted wallpaper and adopts delicate and exquisite European pillar, carve patterns and lines in detail treatment and combination and collocation of large-area warm and cold tones in color treatment.

10

11

12

8 马鞍山万达华府样板间客厅
9 马鞍山万达华府样板间餐厅
10 马鞍山万达华府样板间户型图
11 马鞍山万达华府样板间卧室
12 马鞍山万达华府样板间书房

DEMONSTRATION AREA OF QINGDAO LICANG WANDA PALACE COMMERCIAL STREET
青岛李沧万达华府商业街示范区

开放时间	2013 / 09
建设地点	山东 / 青岛
占地面积	1777 平方米
建筑面积	2332 平方米
OPENED ON	SEPTEMBER / 2013
LOCATION	QINGDAO / SHANDONG PROVINCE
LAND AREA	1,777 m²
FLOOR AREA	2,332 m²

ARCHITECTURE OF COMMERCIAL STREET
商业街建筑

建筑立面风格延续住宅的新古典风格，注重近人的效果，在细节和尺度方面都精心推敲。立面所采用的华贵石材与对称的新古典主义风格相契合。连雨棚、广告位等细节都精心布置，共同打造李沧区"左岸风情街"的宜人景观。

Design of the facade extends the neoclassical style of residence to emphasize the sense of scale, details and scale of the building. Magnificent stones selected for the wall are harmonious with the neoclassical style, and even awning, advertising boards and other details are elaborately arranged to build pleasant landscape of "Left Bank Flavor Lane" in Licang, Qingdao.

1 青岛李沧万达华府商业街动线图
2 青岛李沧万达华府商业街夜景

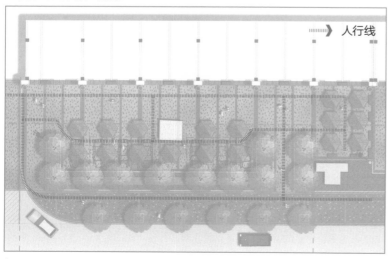

LANDSCAPE OF COMMERCIAL STREET
商业街景观

商业街示范区结合新古典建筑风格，其主题为"花园的早餐——左岸风情街"。植物采用阵列乔木、缓坡地形及馒头型修剪绿篱等设计，形成了丰富的绿植系统；硬质景观设置莫扎特雕塑、绿化式停车位、欧式阳伞桌椅、欧式花钵及移动花箱等园艺小品，营造出活力的、高品位的购物休闲氛围。

The demonstration area of the commercial street of the neoclassical style takes "Garden Prelude: Left Bank Flavor Street" as the theme. Array arbors, gentle slope, dome trimmed hedge and other plants form rich and varied plant system. Hard landscape adopts gardening ornaments, such as Mozart sculpture, greenbelt parking lot, European feature pot and portable flower box to create dynamic and high-grade shopping and entertainment atmosphere.

3 青岛李沧万达华府商业街景观
4 青岛李沧万达华府商业街景观小品
5 青岛李沧万达华府商业街景观绿化
6 青岛李沧万达华府商业街剖面图

DEMONSTRATION AREA OF DONGGUAN CHANG'AN WANDA PALACE
东莞长安万达华府示范区

开放时间	2013 / 06
建设地点	广东 / 东莞
占地面积	3000 平方米
建筑面积	2550 万平方米
OPENED ON	JUNE / 2013
LOCATION	DONGGUAN / GUANGDONG PROVINCE
LAND AREA	3,000 m²
FLOOR AREA	2,550 m²

ARCHITECTURAL PLANNING
建筑规划

简洁方正的体型结合竖向虚实线条的呼应，增强了建筑的整体感。主朝向面整体内凹的门头形式更彰显建筑简洁、大气的整体形象，实现了"小中见大"的设计理念。立面设计中采用了强调竖向线条的新古典式设计手法，优雅的竖向线条元素从建筑的头部至建筑底商中贯通。三段式的新古典式设计横向划分，根据独栋或者两个单元并联，划分位置进行上下浮动，达到整体比例的协调。

The succinct and foursquare shape works in concert with void and solid vertical lines to enhance the sense of integrity of architecture. Inwards concaved door lintel on the frontage display simple but magnificent architectural image, realizing the design concept of "Viewing a World in a Grain of Sand". The facade design adopts the neoclassical design style which emphasizes vertical lines. These elegant lines connect the top of the building with floor stores at the bottom, forming an integrated image. The neoclassical design of three orders is applied to horizontal direction to form irregular space partitioning of one single building or townhouse by two. The overall architectural scale is well coordinated.

1

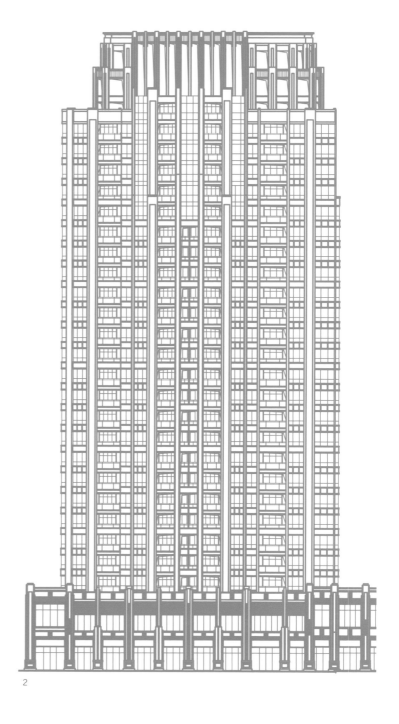

2

1 东莞长安万达华府入口门头
2 东莞长安万达华府建筑立面图
3 东莞长安万达华府商业街小品
4 东莞长安万达华府商业街

DEMONSTRATION AREA OF FOSHAN NANHAI WANDA PALACE
佛山南海万达华府示范区

开放时间	2013 / 08
建设地点	广东 / 佛山
占地面积	3200 平方米
建筑面积	500 平方米
OPENED ON	AUGUST / 2013
LOCATION	FOSHAN / GUANGDONG PROVINCE
LAND AREA	3,200 m²
FLOOR AREA	500 m²

LOBBY OF FINISHED 5A OFFICE BUILDING
实楼甲级写字楼大堂

现代简洁、挺拔大气的风格，笔直向上的线条，加强了空间感。平实的表面，达到稳重、宁静、幽远的效果。木纹石墙面、温和的色泽，让人感到亲切、自然。超高比例的门洞、条形天花，为线性空间平添活力；风口隐藏于其中，显得更加洁净。墙壁上精致细腻的不锈钢条，使巨大体积被分解和柔化，大堂用简洁、单纯的材料组合，给人一种信任感和亲切感。

The lobby is designed in a modern, simple but magnificent style with straight and upward vertical lines to enhance a sense of space. Simple and unadorned surface creates steady and serene effect. Serpeggianto wall with mild color makes people feel warm and comfortable. Ultrahigh and broad doorway and stripe-type ceiling build the linear space more dynamic and vivid. Air vents that are concealed in the ceiling make the space appear to be more clean and tidy. Delicate and exquisite stainless steel strips on the wall disassemble and soften the huge volume. Combination of simple and pure materials endows the lobby senses of trust and intimacy.

1

1 佛山南海万达华府甲级写字楼大堂
2 佛山南海万达华府甲级写字楼电梯间

246 | PART D | THE DEMONSTRATION AREAS OF WANDA SALES PLACES
万达销售实景示范区

3

4

5

6

3 佛山南海万达华府甲级写字楼办公室前台
4 佛山南海万达华府甲级写字楼户型图
5 佛山南海万达华府甲级写字楼办公区
6 佛山南海万达华府甲级写字楼会议室

DEMONSTRATION AREA OF CHANGZHOU WUJIN WANDA PALACE
常州武进万达华府示范区

开放时间	2013 / 12
建设地点	江苏 / 常州
占地面积	2600 平方米
建筑面积	2030 平方米
OPENED ON	DECEMBER / 2013
LOCATION	CHANGZHOU / JIANGSU PROVINCE
LAND AREA	2,600 m²
FLOOR AREA	2,030 m²

ARCHITECTURAL PLANNING
建筑规划

本项目销售物业主要由住宅、SOHO公寓、甲级写字楼和底商组成。住宅依旧延续常州地区 "中吴龙城"之名号，与商业、甲写、酒店的设计概念一脉相承。方案侧重运用经典Art-Deco元素构件，再续"中吴龙城"的神韵，提炼概括出龙鳞的轮廓并结合Art-Deco形式，作为头部装饰构件形成建筑表象，取得推陈出新的效果。

The Properties for Sale of the Changzhou Wujin Wanda Palace are mainly composed of residence, SOHO apartments, 5A office building and floor stores. The residence which continuously inherits the name of "Dragon City" in Changzhou region harmonizes with the design concept of the commercial property, 5A office building and hotel. The classical Art-Deco elements are widely utilized in the design to express the charm of Changzhou, the Center of the Wu State & the Dragon City. Outline of the dragon scale is sketched in the Art-Deco style, serving as decorative element of the door lintel, in an attempt to bring forth a new and creative design.

1 常州武进万达华府动线图
2 常州武进万达华府北入口门头

3

COMMERCIAL STREET
商业街

特色店铺穿插其中,使得每个店铺各有特色,并用其建筑表皮来表达龙鳞正面观看的丰富交错关系,侧面观看时候达到步移"图"异的效果。主色采用浅色系,融入城市环境;基座采用深色优化比例;适度装饰和经典符号的使用,丰富了立面表情,体现了建筑的品质感。

The commercial street is spread with various distinctive and specific stores, expressing rich and vivid image of the facade to make pedestrians enjoy protean and changing "pattern" along with each step. The dominant color of the design adopts light color, making the commercial street harmonious with the urban environment. Dark color is adopted for foundation to optimize scale. Moderate decoration and application of classic symbols enrich the building facade and reflect the quality of the building.

4

3 常州武进万达华府示范区商业街外立面
4 常州武进万达华府示范区商业街夜景
5 常州武进万达华府示范区商业街夜景
6 常州武进万达华府示范区商业街夜景

7
8
9

7 常州武进万达华府示范区景观通道
8 常州武进万达华府示范区入户大堂前景观
9 常州武进万达华府示范区水景
10 常州武进万达华府示范区景观总图

LANDSCAPE
景观

景观设计分为两部分：主次入口景观区、组团景观区。入口景观区运用欧式灯柱、特色精致的铺装和多层次的种植手法，突出典雅大气的形象。组团景观区运用流动的溪流、开阔舒展的草坡、如丝带般的"四季草花"、临水而建的观景平台以及精致典雅的欧式小品，把庄园的浪漫生活氛围和武进江南水乡的情怀表达得淋漓尽致。漫步整个园区，宛若走进欧式经典庄园，可以品读悠长的、华丽舒适的精致生活。

The landscape design of the commercial street can be divided into two parts: the main and secondary entrance landscape area and cluster landscape area. The former adopts the European lamp stands, special and delicate pavement and multilayered plants to highlight elegant and magnificent image while the later utilizes flowing brook, wide lawn, year-round blooming plants, waterside viewing platform and delicate the European ornaments to fully express romantic living environment of the manor and feelings towards Wujin, a southern riverside town in China. Wandering in the garden, one will feel like stepping into a European classical manor to enjoy the gorgeous and comfortable life style.

PROTOTYPE ROOM OF FINISHED PALACE
实楼样板间

住宅样板间运用简单的造型，摒弃多余的装饰，发挥结构本身的形式美。墙面上使用镜面、深色不锈钢、皮革硬包和深木饰面，不同材质的使用产生了质感上的对比，提高了整体的现代感。使用现代简约的家具，在装饰与布置中最大限度地体现空间与家具的整体协调。

Succinct modeling without redundant decoration is adopted to highlight formal beauty of the structure. The wall adopts mirror finish, dark stainless steel finish, leather finish and dark timber finish to reflect distinct comparison in texture, enhancing the modern sense of the whole building. Moreover, the modern and simple furniture reflect overall coordination of space and furniture to the largest extent in decoration and layout design.

11 常州武进万达华府样板间客厅
12 常州武进万达华府样板间主卧
13 常州武进万达华府样板间餐厅
14 常州武进万达华府样板间餐厅陈设

DEMONSTRATION AREA OF WEIFANG WANDA PALACE
潍坊万达华府示范区

开放时间	2013 / 09
建设地点	山东 / 潍坊
占地面积	2500 平方米
建筑面积	2500 平方米
OPENED ON	SEPTEMBER / 2013
LOCATION	WEIFANG / SHANDONG PROVINCE
LAND AREA	2,500 m²
FLOOR AREA	2,500 m²

PROTOTYPE ROOM OF FINISHED LOFT
实楼 LOFT 样板间

LOFT建筑标准层5.4米层高,以55平方米户型为主,采用开间4.6米、进深10.3米的单元组合拼接。样板间利用高挑空间,将其纵向一分为二,上下两层由室内楼梯连通,打造开敞与私密空间分割得当的1厨2卫1房2厅格局,给空间纳入更多层次。再配以不同质地、不同属性的装饰材料,营造出温馨、舒适的居住氛围,将功能与风格巧妙合一。

The typical floor height of the LOFT is 5.4 meters and area of the main house type of it is 55 square meters with width of 4.6 meters and depth of 10.3 meters. By virtue of the high space, the prototype room is divided into two floors, connecting by interior stair to create open and private spaces of appropriate layout with one kitchen, two bathrooms, one bedroom and two living rooms, which endowing the space with rich spatial levels. The finishing materials in various textures and properties build warm and comfortable living space with full functions and delicate style.

1

1 潍坊万达华府LOFT一层书房
2 潍坊万达华府LOFT客厅

5

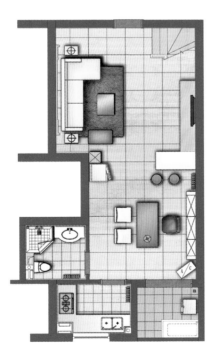 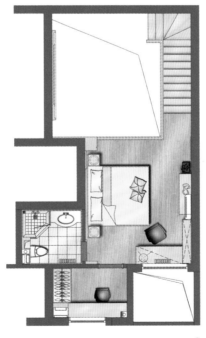

在空间布局上：一层为客厅、餐厅、厨房和卫生间，楼梯的设计采用现代风格；二层空间为卧室，并兼顾书房的功能。整体空间采用温润的木色系，家具、摆设都同样走轻盈自然路线。

The spatial layout: the 1F is set with living room, dining room, kitchen and bathroom and the stair is designed in modern style; the 2F is arranged with one bedroom which can also serve as a study room. The whole space is finished in mild wood color, and the furniture and the furnishings adopt similarly natural and casual design style.

6

3 潍坊万达华府LOFT客厅
4 潍坊万达华府LOFT一层书房
5 潍坊万达华府LOFT二层卧室
6 潍坊万达华府LOFT户型图

DEMONSTRATION AREA OF ANHUI BENGBU WANDA PALACE COMMERCIAL STREET

安徽蚌埠万达华府商业街示范区

开放时间	2013 / 04
建设地点	安徽 / 蚌埠
占地面积	2900 平方米
建筑面积	2200 平方米
OPENED ON	APRIL / 2013
LOCATION	BENGBU / ANHUI PROVINCE
LAND AREA	2,900 m²
FLOOR AREA	2,200 m²

1

2

COMMERCIAL STREET
商业街

安徽蚌埠万达华府商业街特点可用两个字形容:"红"和"变"。红色装饰飘板仿佛红色飘带,如行云流水般游走于建筑立面之上,贯穿始终,活力四射,灵动、闪耀。建筑立面通过错落有致的造型、多样性的材料、变幻色彩的组合应用,契合设计主题,迎合潮流,张扬个性,极富现代感。通过挑板外檐LED线条灯光的延续,将整个底商形成一个整体,局部大面积采用导光板,挑出建筑立面,运用先进的控制系统,LED导光玻璃同步变色,从而形成跃动的风景线和浪漫的光影造型,使静止的形态焕发出生机,调动着人们意识深处的主观情绪,产生情感的共鸣。转角处采用投光照明,使其形成一个有层次渐变的受光面,在光环境下突出建筑的肌理和建筑的几何空间。商铺雨棚采用筒灯,增加路面的照度,使人流更清楚的行进路线。

The feature of commercial street of Anhui Bengbu Wanda Palace can be summarized and described in two Chinese characters: "Red" and "Variety". The red decorative overhanging plates float and flow upon building facades throughout the commercial street just like red ribbons, dynamic, vivid and sparking. The facade design adopts abundant and well-organized modeling, various materials and color schemes to echo with the design motif and trend, expressing the characteristics and reflecting modern quality of the design. The LED linear lighting on outer edge of the overhanging plates integrates the floor stores as a whole. Light guide plates, which are widely utilized at some places, highlight the building facade. Advanced control system is adopted to realize simultaneous color changing of LED guide glasses and create dynamic scenery and romantic lighting modeling, making static forms reflect dynamic and enabling pedestrian to generate emotional resonance. The flood lighting is set at corners to form gradually changing and multileveled reflective surface, highlighting texture and geometric space of the building in luminous environment. Down lamps which are adopted for awnings of the stores for road lighting provide bright and clear route guiding for pedestrians.

1 安徽蚌埠万达华府商业街示范区效果图
2 安徽蚌埠万达华府商业街示范区夜景
3 安徽蚌埠万达华府商业街示范区外立面效果图
4 安徽蚌埠万达华府商业街示范区建筑外立面

PART **E**

WANDA COMMERCIAL PLANNING
**CLASSIC PROJECTS
RETROSPECT OF WANDA
PROPERTIES FOR SALE**
万达销售物业经典项目回顾

REPRODUCTION OF TRADITION: RETROSPECT OF CLASSIC PROJECTS OF THE CHINESE-STYLE COURTYARDS OF WANDA ORGANIC AGRICULTURE GARDEN

古为今用再造传统
——万达有机农业园中式四合院经典项目回顾

文／万达商业地产设计中心总经理　尹强

万达有机农业园位于北京市延庆县香营乡，占地387公顷，地势平坦，四周群山环抱。管理用房位于农业园中心位置，北倚燕山，南有人工湖，场地自然条件优越，周边村落建筑多为灰色砖瓦民房。怎样使一栋建筑能够在满足使用功能的前提下，最大限度地表现出对这一片依山傍水的优美自然环境的尊重与融合呢？最终，管理用房被设计为一栋单层钢筋混凝土框架的中式四合院建筑，建筑面积1680平方米，占地3380平方米，庭院面积1700平方米。

将传统建筑文化置于现代建筑功能背景下进行再创作，自始至终是我们认真思考的问题。因此，"再造传统"成为了创作的核心，从功能布局、空间处理、形式与功能的结合到景观、室内设计等方面，无不体现出对传统建筑文化的尊重、理解、演绎和再造。

Located in Xiangying Town, Yanqing County, Beijing, Wanda Organic Agricultural Garden covers an area of 387 hectares of flat land surrounded by mountains. The management office is in the middle of the Agricultural Garden, connected to Yanshan Mountain in the north and an artificial lake in the south, enjoying excellent natural conditions. The surrounding villages are mostly houses in gray bricks. How to show respect to and integrate with this beautiful environment to the most extent for a building under the premise of meeting use function? Finally, the management office is designed to be a Chinese-style courtyard in single reinforced concrete frame, covering the floor area of 1,680 square meters, the land area of 3,380 square meters and courtyard area of 1,700 square meters.

Wanda has always been thinking of the issue of recreating the traditional architectural culture under the background of modern building function. So, "recreation of tradition" becomes the core of creation, reflecting

（图1）迎宾庭院

一、功能布局的传统再造

管理用房被设计为典型的北方四合院风格的建筑，功能上分为公共活动区、居住客房区及辅助用房区三部分。以会客厅等公共接待部分构成院落的中心建筑，其他功能用房环绕在其周围。借用传统四合院的形式，创造性地围合成"迎宾院"、"玉兰堂"及"春"、"夏"、"秋"、"冬"共计六个大小不一的院落。

respect, understanding, evolution and reproduction for traditional architectural culture in aspects of functional layout, space treatment, combination of form and function, landscape and indoor design, etc.

I. RECREATION OF TRADITIONAL FUNCTION LAYOUT

The management office is designed into a typical

（图2）迎宾庭院

（图3）玉兰堂

1. 第一进院落

第一进院落为迎宾庭院，是泊车落客的场所。为了突出建筑群的庄重、大气，院门位置突破了四合院原有格局，将院门开在中轴线上，院门也没有采用民居形式，而是选用王府大门级别的院门，以此烘托隆重气氛。在厅堂位置设置的会客厅前设计了可以进入车辆的门廊，使得宾客可以便捷地进入室内，在门廊的形式上选用了高于民居规制的卷棚歇山屋顶，用以突显主立面的华丽和庄严。在会客厅前设计了抄手游廊，丰富了空间层次，增加了会客厅的私密性（图1、图2）。

2. 第二进院落

第二进院落为玉兰堂，是本建筑群中最具有北方四合院风格特点的主要庭院。建筑尺度、细节及建筑形式严格遵守传统模数做法，并恪守主从关系，两间套房的起居部分位于坐北朝南的正房，端庄大气；卧室部分位于带跨院的"耳房"私密、安静。四套客房位于东西厢房（图3）。

3. 东西两边的院落

为满足卧房的使用功能，创造性地在东西两边所有卧房南边对应地设置了四个精致的内院。为了保证卧房的私密性，在抄手游廊靠内院一侧改为实墙与中心庭院玉兰堂分开，墙上开什锦花窗，相互借景。四段抄手游廊利用大玻璃窗封闭处理，围绕中心庭院形成围合，贯通室内各部分的交通，满足了现代功能的要求。

二、空间处理的传统再造

在建筑空间处理上，有继承有创新——遵从传统四合院建筑的主从关系，主要院落空间沿轴线布置，依次展开，层层递进；次要庭院布置在东西两边，主从有序。

由于东西厢房进深较大，若按单坡形式，屋脊高度将超过正房，为此将屋顶设计成勾连搭形式，有效降低了屋脊高度。堂皇的院门、高大的会客厅（厅堂）及庄重的（正房）主人套房这一组最重要的建筑位于中轴线上。这组建筑规格最高、体量最大、屋脊最高，以此突出正位为尊的传统象征意义。

courtyard in the northern part and functionally divided into three parts of public activity area, residential guestroom area and auxiliary room area. The reception room and other public reception parts constitute the central buildings of courtyard surrounded by other functional rooms. Six courtyards in different sizes are innovatively created in the type of the traditional courtyard, including "Guest-greeting Yard", "Magnolia Yard", "Spring", "Summer", "Autumn" and "Winter" Yards.

1. THE FIRST COURTYARD

The first is Guest-greeting Yard, the place for guests to get off and parking. To emphasize solemn and magnificent atmosphere of buildings, the gate is designed to break the original pattern of courtyard and is placed in the middle axis. The gate in Palace level is chosen to display solemn atmosphere rather than dwelling type. The stoep is designed in front of the reception room in the hall for guests' convenience to enter into the room. The style of stoep is hip-and-gable roof which is better than vernacular dwelling to highlight magnificence and solemnity of the main façade. The veranda is designed in front of the reception room, enriching space layers and increasing privacy of the reception room (Figure 1 and Figure 2).

2. THE SECOND COURTYARD

The second is Magnolia Yard and it is the main courtyard in the buildings with the typical characteristics of courtyard in the northern part. The yard is in strict compliance with traditional style for architectural scale, details and architectural shape. It also follows traditional relationships between the primary building and the secondary ones that living parts for two suites are in the principal room facing south, which is dignified and magnificent and the bedroom is located in the aisle with the lateral courtyard, which is private and tranquil. Four guestrooms are wing-rooms in the east and west sides (Figure 3).

3. COURTYARDS IN THE EAST AND WEST SIDES

Four delicate internal yards are correspondingly placed in the south of bedrooms in the east and west sides to satisfy use function of bedrooms. To guarantee privacy of the bedrooms, dead wall is used to separate with Magnolia Yard at the side of the veranda near to the internal yard. Lattice window is in the wall for mutual view. Four verandas are circling the central courtyard with big glass window, connecting all paths towards rooms and meeting modern function requirement.

II. RECREATION OF THE TRADITIONAL SPACE

Innovation is integrated in the inheritance for treatment of architectural space. The architectural space firstly

三、传统形式与现代功能的有机结合

1. 建筑形式
本项目为钢筋混凝土框架结构与传统木作建筑形式的结合。外露柱子采用钢筋混凝土圆柱,外露的框架梁也刻意模仿木作的梁、板、枋形式;屋面形式也按照木屋架屋面曲线进行支模浇筑;其他传统外露构件则完全采用木作;最后,所有外饰均采用传统油漆彩画工艺——包浆、裹麻、卓彩,从而达到了建筑外观与传统木作形式上的一致(图4)。

2. 功能与形式的结合
会客厅内部功能需要三开间,而外立面应采用五开间形式。在处理好内部梁枋衔接后,创造性地采用两套柱网,既满足了使用功能又保持了传统建筑形式。两边厢房按传统应为东西向布置,为了既满足传统制式需要又能达到现代功能的舒适要求,设计将两边厢房分别改为四间南向客房,同时将朝向主院一侧的山墙用设置假窗的办法,做成了厢房的式样,化解了矛盾。

3. 门窗形式
采用实木复合中空玻璃门窗,形式上模仿传统门窗形式的比例,在木窗棂内侧绷纱窗,从而解决了野外防蚊蝇的实际需要。

四、景观设计的传统再造

努力再现传统建筑景观场景,同时结合现代使用功能及场所空间尺度,在景观设计时有所创新、有所突破。

1. 门前广场
为利用田野进入建筑群的过渡空间,广场两侧设置了

(图4)游廊

follows hypotactic relationship of traditional courtyard building that the principal courtyards are placed in the axis successively and progressively and subordinated courtyards are placed in the east and west sides.

Due to large depth for the wing-rooms in the east and west sides, ridge will be higher than that of the central room if single slope is adopted. Therefore, the roof is designed in connection form to effectively reduce ridge height. The most important rooms are located in the central axis, including the magnificent gate, the grand reception hall (living room) and the master's solemn suite. The highest system, biggest volume and highest ridge of these rooms highlight the traditional symbolic significance of respecting the greatest power.

III. THE APPROPRIATE COMBINATION OF THE TRADITIONAL STYLE WITH THE MODERN FUNCTIONS

1. ARCHITECTURAL STYLE
The project is a combination of reinforced concrete frame with the traditional wooden building. The external pillars are circular reinforced concrete column and the framed girder also deliberately simulates wooden girder, board and square-column. Formwork pouring is made for roof according to curve of wooden roof and other external traditional components are all wooden. Finally, all external decorations are in traditional color painting process: wrapped slurry, wrapped in linen and paint-realizing the consistency between architectural appearance and traditional wooden type (Figure 4).

2. COMBINATION OF FUNCTION AND STYLE
The internal function of the reception room requires three bays while the external façade five bays. Two sets of innovative column grid meet use function and keep the traditional architectural type after the connection between interior girder and square column is well handled. The wing-rooms are traditionally distributed in the east and west sides. To meet the traditional system and modern comfortable requirement, the wing-rooms at both sides are changed into four guestrooms facing the south. At the same time, the gable facing the central courtyard is changed into a wing-room by setting faked window to resolve the contradiction.

3. DOOR AND WINDOW TYPE
The doors and windows are made with solid wood installed composite and hollow glass to simulate traditional door-window proportion. The screen inside window lattice protects from mosquitoes and flies.

(图5)春园

(图6)夏园

(图7)秋园

(图8)冬园

成排的栓马桩、下马石，高大的石狮彰显出入口空间的威严。广场两侧选用低矮的树木借以衬托院门的高大。

2. 迎宾庭院

青石板漫地，浮雕居中。中间御道形式的铺砌强调了轴线的重要，两侧成行的银杏树及石雕灯柱强调了环境的秩序。

3. 玉兰堂

正房前种植两颗高大的玉兰，喻示"金玉满堂"的美好寓意；院内还种植了海棠、石榴、牡丹等寓意美好的植物。为满足商务活动需要，将青石十字甬道分割出的四块草地进行了硬化，采用青砖竖砌的方式保持了四合院庭院的形式。

4. 春夏秋冬四个庭院

以海棠、梧桐、红枫、腊梅体现四季景观的典型树种，营造东西两边的四个庭院（图5~图8）。

五、室内设计的传统再造

采用中式风格，结合现代功能需要对传统建筑内饰进行适当调整和创新，使之更能符合现代人们的审美要求。为保证重要场所的氛围，在家具选择上也尽可能保持原汁原味——例如客厅就完全采用条案、八仙桌等传统家具形式。

采用旋子彩画中的墨线大点金做法对室内空间进行彩绘装饰，其中会客厅采用了等级更高的金线大点金做法。天花图案选用"玉堂富贵"图案（图9、图10）。

万达有机农业园管理用房的设计从满足使用功能出发，在尊重传统建筑形式的同时又有所创新，不愧为现代使用功能与传统建筑形式完美结合的成功范例。

IV. RECREATION OF THE TRADITIONAL LANDSCAPE DESIGN

1. SQUARE IN FRONT OF THE GATE

The square is a transition space from the outer field to the buildings. Horse-fastening pile and stone in rank are placed both sides of the square. Tall stone lion demonstrates dignity of the entrance space. Low trees are chosen on both sides of the square to set off high gate.

2. GUEST-GREETING YARD

The yard is paved with blue flagstone with emboss in the middle. The royal road in the middle emphasizes importance of axis and gingko trees and stone sculpture lamp posts on both sides highlight orderly environment.

3. MAGNOLIA YARD

Two high magnolia trees in front of the central building signify beautiful meaning of "abundant wealth". Also other plants signifying good meanings are planted in the yard, like begonia, guava and peony. To meet requirements of commercial activities, four pieces of grassland divided by cross-shaped aisle are hardened by paving with black bricks, keeping the courtyard type.

4. SPRING, SUMMER, AUTUMN AND WINTER YARDS

Four yards are created in the east and west sides with the typical trees in four seasons including begonia, Chinese parasol, red maple and wintersweet (Figure 5 to Figure 8).

V. RECREATION OF THE TRADITIONAL INTERIOR DESIGN

Combined with modern functional demand, the Chinese style is adopted for proper adjustment and innovation for the interior decoration of the traditional building, making it fit for modern aesthetic requirements. To keep the atmosphere of important places, furniture should stay with the original style. For example, in the living room, it would be appropriate to use console table, square table, etc.

Gold pointing with inked line in tangent circle pattern is adopted for colored drawing decoration in the indoor room, among which the gold pointing with golden line in a higher level is applied in the reception room. The ceiling is in patterns of magnolia, begonia and peony (Figure 9 and Figure 10).

Starting from use function, the design of management room of Wanda Organic Agricultural Garden is innovating while respecting the traditional building. It is deserved to be the successful case of perfect combination of modern use function with the traditional building style.

（图9）起居室

（图10）大客厅彩绘

WANDA CHINESE-STYLE COURTYARD
万达中式四合院

入伙时间	2012 / 10
建设地点	北京 / 延庆
占地面积	3375 平方米
建筑面积	1680 平方米
ADMISSION TIME	OCTOBER / 2012
LOCATION	YANQING COUNTY / BEIJING
LAND AREA	3,375 m²
FLOOR AREA	1,680 m²

ARCHITECTURAL PLAN
建筑规划

设计为典型的北方四合院风格的建筑，功能上分为公共活动区、居住客房区及辅助用房区三部分。建筑空间处理上，首先遵从传统四合院建筑的主从关系，主要院落空间沿轴线布置，依次展开，层层递进，有继承有创新。

次要庭院布置在东西两边，主从有序。借用传统四合院的形式，创造性地围合成"迎宾庭院"、"玉兰堂"及"春"、"夏"、"秋"、"冬"共计六个大小不一的院落。建筑形式为钢筋混凝土框架结构与传统木作建筑形式的结合。传统外露构件完全采用木作，所有外饰均采用传统彩画工艺，从而达到了建筑外观与传统木作形式上的一致。

1

1 清风明月堂景观
2 云纹图案、砖雕牡丹

The typical courtyard building in the northern part is functionally divided into three parts including public activity area, guestroom area and auxiliary room area. The architectural space follows hypotactic relationship of the traditional courtyard building that the principal courtyards are placed in the axis successively and progressively. Innovation is integrated in the inheritance of the courtyard.

Subordinated courtyards are distributed to the east and west sides and it is orderly for the principal and subordinated courtyards. Six courtyards in different sizes are creatively built in the type of traditional courtyard, including Guest-greeting Yard, Magnolia Yard and Spring, Summer, Autumn and Winter Yards. The building is a combination of reinforced concrete frame structure with the traditional wooden architecture that all external traditional components are wooden and external decoration is in the traditional color painting process, thus to reach consistence of architectural appearance with the traditional wooden type.

3 清风明月堂门厅
4 清风明月堂前院

5

6

5 玉澜堂景观
6 清风明月堂连廊
7 步秋山房
8 步秋山房立面图

LANDSCAPE
景观

景观设计旨在为业主营造具有深厚文化内涵和品位的室外活动空间。努力再现传统建筑景观场景，同时结合现代使用功能及场所空间尺度在景观设计时有所创新有所突破。正房前种植两颗高大的玉兰，喻示"金玉满堂"的美好寓意，院内还种植了海棠、石榴、牡丹等寓意美好的植物。以海棠、梧桐、红枫、腊梅体现四季景观的典型树种营造东西两边的"春"、"夏"、"秋"、"冬"四个庭院。

The landscape is aimed to create outdoor activity space with profound cultural content and taste for owners. It is strived to reproduce landscape of traditional building and innovate and breakthrough in designing landscape combined with modern use function and space dimension. Two high magnolia trees in front of the central building signify beautiful meaning of "abundant wealth". Also other plants signifying good meanings are planted in the yard, including begonia, guava and peony. Four yards of "Spring", "Summer", "Autumn" and "Winter" are created in the east and west sides with typical trees in four seasons including begonia, Chinese parasol, red maple and wintersweet.

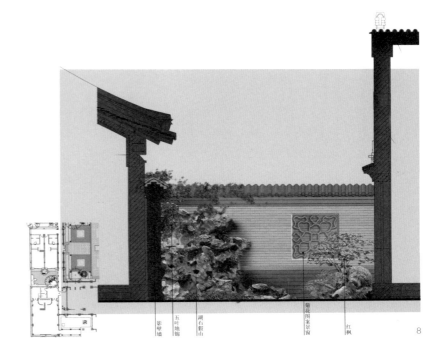

9

9 三友小筑立面图
10 梧竹幽居
11 三友小筑

10

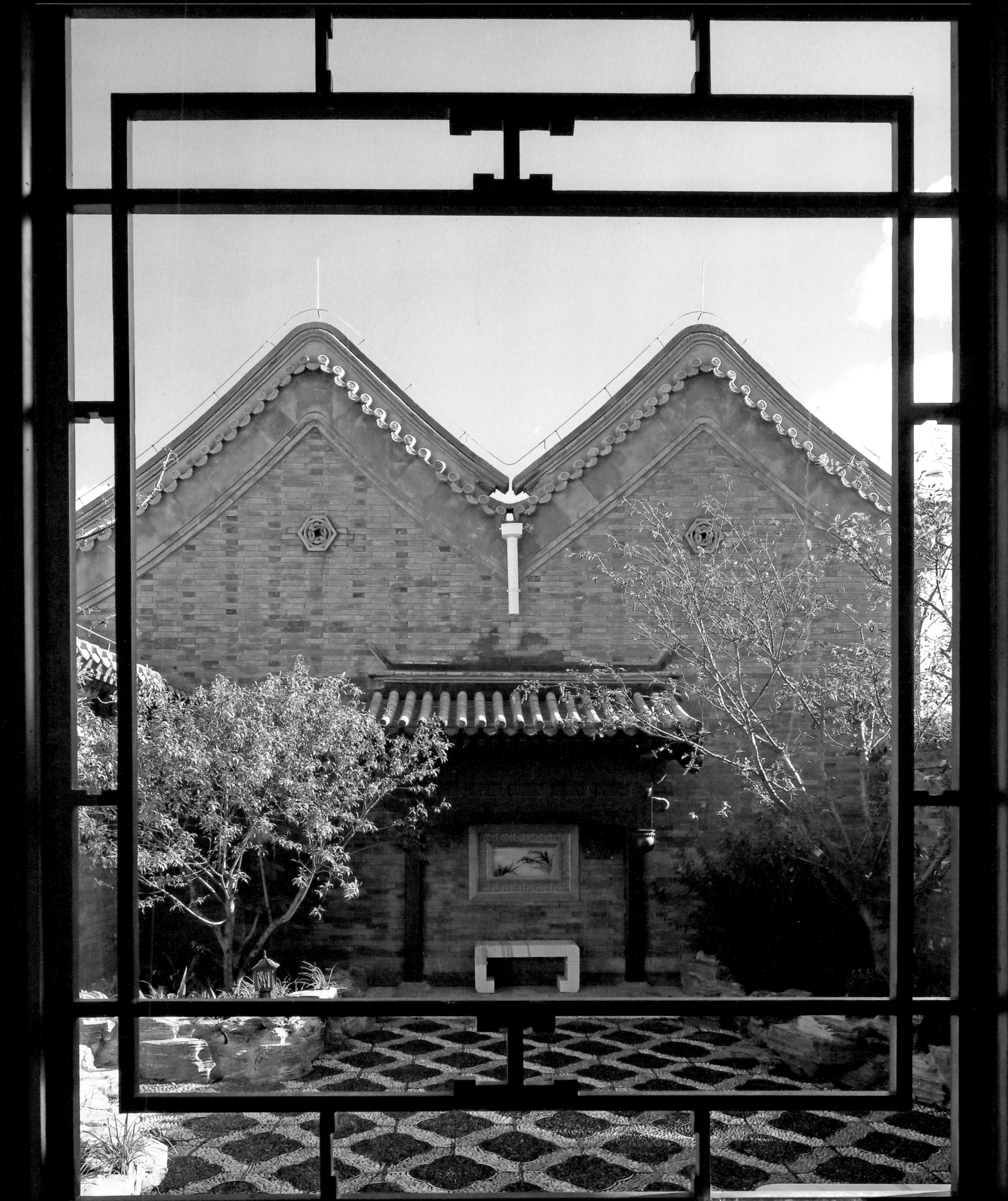

INTERIOR DECORATION
内装

室内采用中式风格,结合现代功能,对传统建筑内饰进行适当调整和创新,使之更能符合现代人们的审美要求。为保证重要场所的氛围,在家具选择上也尽可能保持原汁原味的传统风格,例如客厅就完全采用条案、八仙桌等传统家具形式。室内空间采用旋子彩画中的墨线大点金做法进行彩绘装饰,其中会客厅采用了等级更高的金线大点金做法。天花图案选用了"玉堂富贵"图案。

Combined with modern functional demand, the Chinese style is adopted for proper adjustment and innovation for indoor decoration of the traditional building, making it fit for modern aesthetic requirements. To keep the atmosphere of important places, natural-traditional furniture should be kept and used in the living room, including console table, square table, etc. The gold pointing with inked line in tangent circle pattern is adopted for colored drawing decoration in the indoor room, among which the gold pointing with golden line in a higher level is applied in the reception room. The ceiling is in patterns of magnolia, begonia and peony.

12 套房卧室
13 套房会客厅

PART F

WANDA COMMERCIAL PLANNING
EPILOGUE
后续

RETROSPECT OF WANDA PROPERTIES FOR SALE
万达销售物业历程回顾

文／大连万达商业地产股份有限公司境外地产中心副总经理
兼项目管理部总经理　刘拥军

万达自1988年创立后，随着对商业地产的不断研究，产品一代店逐渐发展为城市综合体。这期间，作为万达设计管理的核心团队——万达商业规划研究院起到了全面的管理作用，销售物业并没有单独的设计管理部门（图1）。

（图1）第一代万达：长沙解放军西路万达

随着万达商业地产的不断发展壮大，这种复合型业态的管控逐渐进入更加精细化的阶段，面向集团自身的持有型大商业和面向市场的销售物业之间的管控目标与管理手段形成了较大的差异。为了更好地对销售型物业进行精细化设计管理，2008年10月，集团决定成立住宅设计部。

住宅设计部成立之初只有3人，并且只有建筑一个专业的人员，没有明确的管控职责，没有成型的管理工具，甚至没有独立的办公空间，却立刻负责起了2009年度苏州平江、上海周浦、青岛CBD、重庆南坪、西安民乐园、沈阳太原街、南京建邺、洛阳涧西等八个当年开业的以及在建的十几个万达广场的销售物业设计管控。住宅设计部不负众望，不仅让每个项目的户型设计得到市场认可，也形成了《销售物业技术管理细则》等制度文件，使销售物业管理走上正轨（图2）。

2009年住宅设计部正式纳入项目管理中心，以市场

Since its founding in 1988, Wanda has been developing the first generation store into city complex along with persistent research for the Commercial Properties. During this time period, as the core team of Wanda design management, Wanda Commercial Planning & Research Institute had played a comprehensive management role as the Properties for Sale did not have its own management department (Figure 1).

With development and growth of Wanda Commercial Properties, the control & management of this enterprise complex with multiple industries went to a refining stage and its objectives and measurements appear huge control differences between Group-oriented Properties for Holding and market-oriented Properties for Sale. To better design and manage the Properties for Sale, the Group decided to set up Residence Design Department in October 2008.

At the beginning, there was only one architectural specialty, totaling three employees in the Residence Design Department without clear assignment of responsibilities nor separate office space, but they did a great job in design management of Wanda's Properties for Sale including eight Wanda Plazas opened in that year (Pingjiang in Suzhou, Zhoupu in Shanghai, Qingdao CBD, Nanping in Chongqing, Minle Garden in Xi'an, Taiyuan Street in Shenyang, Jianye in Nanjing and Jianxi in Luoyang) and a dozen of on-going Wanda Plaza construction projects. Beyond the expectation, the Residence Design Department not only makes the market accept the housing type design in each project, but also formulates management system like *Design and Control Manual of Properties for Sale*, making the management of the Properties for Sale get on the right track (Figure 2).

In 2009, the Residence Design Department was officially merged into the Project Management Center, establishing market-oriented and client-focused guiding ideology of the products. The Project Management Center grasps the following features:

1. DESIGN RHYTHM

Wanda Properties for Sale are cash flow products and

（图2）大连东港公馆小区景观

（图3）东莞东城万达公馆售楼处建筑

（图4）烟台芝罘万达公馆售楼处景观

（图5）东莞厚街万达公馆售楼处内装

为导向，以客户为中心的产品指导思想确立。项目中心设计部抓住了以下特点：

1. 设计节奏

万达销售物业是现金流产品，抓住市场适应面宽的主流产品，快速出货，是销售物业产品定位与设计的关键所在，快速高效的设计节奏是实现快速销售的前提保障。

2. 标准化设计

由于万达广场在各地的位置相似性，总体规划的模块化，住宅、SOHU、写字楼、商铺、售楼处的产品具备可标准化的所有条件。

基于以上特点，项目管理中心设计部在2010年至2012年，用了两年的时间，将住宅的A、B、C、D版，B版写字楼、D版公寓、各类商铺及A、B版售楼处等均完成了可限额设计的、全部版本全专业的标准化施工图。住宅B、C版和公寓D版还完成了实景样板基地。万达销售物业的设计与建设不仅快速、高效、成本可控，同时有了标杆样板，使销售物业的品质上了一个大的台阶。（图3~图5）

2012年的10月，项目管理中心设计部实现了全专业的管控，对各地项目公司设计部的培训与管理实现了月度考核与培训相结合的管理机制，第三方审图咨询机构的应用，卖场日报制度、品质周报制度、设计部项目经理负责制等管理动作，控制住了项目的设计进度、品质、质量和成本。

2013年度，随着万达集团稳坐全球第二大不动产企业、大力扩展文化旅游产业、跨国发展取得重大突破的进步，全年施工面积5179万平方米。万达一家企业的商业地产施工面积接近整个欧洲商业地产施工面积的总和。企业的发展，对销售物业的设计管理提出了新的要求，集团将原来的项目管理中心设计部独立出来，成立设计中心，人数扩编为140人，为下一步万达销售物业的产品设计提供了强大的组织保证。我们坚信设计中心能够肩负起重任，为实现集团的战略目标做出更大贡献。

mainstream products adapting to the market. Fast delivery is the key of product positioning and design of the Properties for Sale. Fast and efficient design rhythm is the precondition and guarantee of achieving fast sales.

2. DESIGN STANDARDIZATION

Due to similarity of locations of Wanda Plazas in various regions, the overall planning can be modeled, and products from residences, SOHUs, office buildings, stores and sales offices contain all conditions of standardization.

Basing on the above features, the Design Department of Project Management Center spent two years achieving construction drawing standardization of quotable design in all types and specialties of residential types of A, B, C and D, office buildings of type B, apartments of type D, various stores, sales offices of type A and B from 2010 to 2012. Beside, residence of types B and C and apartment of type D completed realistic prototype base. The design of Wanda Properties for Sale is not only fast, efficient and cost-controllable, but also making the quality of the Properties for Sale reach a new level through benchmarking the prototype room (Figure 3 to Figure 5).

In October 2012, the Design Department of Project Management Center achieved the goal of controlling and managing in all specialties and implemented the management system of combination of monthly assessment and training for every project company design department. The management tools, such as application of drawing-checking and consultative institute of the third party, daily report system of sales place, weekly report system of quality and responsibility system of project manager in the Design Center, have controlled project design progress, features, quality, and cost as well.

In 2013, as Wanda stably ranks the second largest real estate business of the world and expands its business to the cultural tourism industry and multinational development with huge success, the construction area of Wanda in the whole year reached 51.79 million square meters. The construction area of the Commercial Properties of Wanda Group was close to the whole construction area of commercial properties in Europe. The development of the enterprise put forward new requirements for design management of the Properties for Sale. Therefore, Wanda Group turned the original Design Department of Project Management Center into the Design Center, increasing staff to 140 members to guarantee the design quality of products of Wanda Properties for Sale in the next step. It is undoubted that the Design Center is capable of shouldering the responsibilities and making greater contributions for fulfilling Wanda's strategic objectives.

RELATIONSHIP BETWEEN WANDA PROPERTIES FOR SALE AND WANDA PLAZAS: "AGGREGATED VALUE" IN THE RESOURCE SYMBIOSIS ERA

万达销售物业与万达广场的关系
——资源共生时代下的"聚合增值"

文／万达商业地产设计中心南区设计部总经理　张东光

一个快速发展的社会，需要聚合增值、资源共生的地产互生模式，融入缤纷的生活资源，全面将生活配套纳入其中，以满足日益多元化的生活需求，形成一种有机互动的交流平台。因此，万达广场是一个时代发展的必然趋势，是驱动现代化城市发展的引擎。

在此宏观时代背景下，万达集团的成功之道既是一种商业传奇，亦归为一种商业宿命。万达集团做对迅速勃发的程序、找到不断成长的密码——不断定义、代表和诠释中国商业地产发展的方向和市场契合的标准，使销售物业与万达广场达到高度统一。

万达广场是万达独创的商业地产模式，包括大型商业中心、室外商业街、五星级酒店、写字楼及公寓等。万达广场构筑了都市中的居住、办公、商务、购物、文化娱乐、社交、游憩等各类功能，是这些功能相互复合、相互作用、互为价值链的高度集约的大型独立商圈。

万达广场打造的城市中心是感受繁华、出行便捷、生活便利，并且物业具备保值增值功能、宜商宜住宜投资的优良物业（图1、图2）。

A rapid development society requires property integrating into wonderful life resources and containing supporting living facilities, generating aggregated value in the symbiotic resource era, thus to meet diversified living demands and form an interactive exchange platform. Therefore, Wanda Plaza is inevitable development tendency of the times and also the driving force for development of modern city.

Under the background of macro era, the success of Wanda Group is a business miracle and also a business destiny. Wanda Group has found out the growth password for the rapid expanding process. Wanda Group has continuously defined, represented and explained the matching standard between development direction of the commercial property and the market in China, achieving high unity between the Properties for Sale and Wanda Plaza.

Wanda Plaza is the original commercial property mode of Wanda Group, including large-scale business center, outdoor commercial pedestrian streets, five-star hotels, office buildings and apartments, etc. Wanda Plaza integrates various functions in the city including residence, office, business, shopping, culture and entertainment, social contact and travel and it is a large-scale independent intensive business area. These

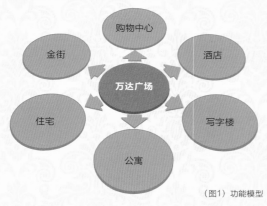

（图1）功能模型

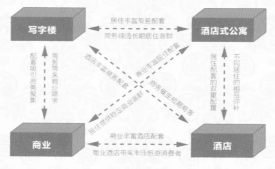

（图2）功能价值体系

销售物业指万达广场中可销售部分的物业类型，与自持物业共同组成万达广场，包括：写字楼、公寓、室外商业街、住宅、商铺、销售商务酒店、销售酒楼及车位等。

销售物业是万达广场中必要的组成部分，从开发、运营和设计角度来看都是不可或缺。

1. 开发角度
改革开放后中国商业形态发展经历三个发展阶段，

functions recombine and interact in the area as value chain of each other.

Wanda Plaza in the urban area is full of prosperity, convenient in travel and living. The property is excellent property having the functions of maintenance and appreciation of values and fit for business, residence or investment (Figure 1 and Figure 2).

The Properties for Sale means the Properties for Sale in Wanda Plaza, together with the Properties for Holding to constitute Wanda Plaza, including office buildings,

由早期自发形成的沿街商铺向特色主题商业街转型，如今已发展成城市万达广场形式，未来正逐步结合文化衍生出更高端的形态。万达广场的发展也伴随历经了几个阶段（图3）。

（图3）万达商业综合体发展阶段

万达第一代单店、第二代组合店：其功能简单，主要为纯商业或商业加居住。为了确保项目开发现金流，商业大量出售，带来了诸多问题：
◆ 后期统一经营难以保证；
◆ 自持比例过低，无法形成商业圈；
◆ 商铺售价过高，散铺投资收益预期无法实现；
◆ 租金收入无法形成后续开发动力。

2006年宁波鄞州万达广场开业，产生了第三代万达广场：步行街加主力店的基础搭配，写字楼、住宅、公寓、底商的辅助搭配；更为核心的区别是：与一代、二代店相比，新建的商业中心不再出售，全部留作持有型物业，只以租的形式，实现统一管理，形成"以售养租"的商业投资模型。销售物业回笼庞大的资金流，支撑着万达广场的开发建设，同时也为万达的迅速扩张提供保障。

万达广场的商业氛围和生活配套是销售物业的王牌，是其他楼盘无法追赶的鸿沟，凭借它得天独厚的优势，创造了销售奇迹，平稳地渡过一个个房地产调控期。

在开发过程中，销售物业为万达广场提供造血功能，万达广场又为销售物业造势，营造良好的口碑和生

apartments, outdoor commercial streets, residence, stores, sales business hotels, sales restaurants and parking, etc.

The Properties for Sale are a necessary part in Wanda Plaza and indispensible from the perspectives of development, operation and design.

1. DEVELOPMENT PERSPECTIVE
After the reform and opening up of China, the business types in China have undergone three development stages. One is transforming from spontaneous stores along the streets at early stage to themed business street. Now it is developing into the Wanda Plaza type in the city and deriving into a higher-end type gradually combined with culture in the future. The development of Wanda Plaza has also experienced several stages (Figure 3).

of Wanda and the second combined store: having simple function, mostly for business or business plus residence. To ensure cash flow for project development, a large amount of properties is for sales followed by many problems:
◆ Difficult for unified management at the later stages;
◆ Difficult to form a business circle because of low holding proportion;
◆ Unrealized expected investment benefit for over-high sales price of stores;
◆ Income from rent cannot support follow-up development.

Yinzhou Wanda Plaza opened in Ningbo in 2006 generates Wanda Plaza in the third generation: the fundamental match of pedestrian streets and principal stores and auxiliary match of office buildings, residences, apartments and floor stores. The core difference is: comparing with stores in the first and second generations, the new commercial center would be kept as holding properties for rent rather than for sales. Unified management can be realized to constitute commercial investment mode of "supply rent with sales". The huge cash flow returned from Properties for Sale supports development and construction of Wanda Plaza and also guarantees rapid expansion of Wanda Group.

The commercial atmosphere and supporting living facilities of Wanda Plaza is the trump of the Properties for Sale and also the wide gap cannot be chased by other real estates. Upon the unique advantages, Wanda Plaza has created sales miracle and smoothly gone through real estate control periods.

The Properties for Sale supply capital resource

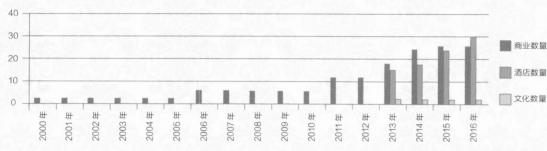

（图4）万达开业项目统计表

活配套。两者相互补充，相互依存，实现产品链与资金链的优化组合和良性循环（图4）。

2. 运营角度

万达广场的运营需要人气，销售物业中写字楼的办公人群、公寓及住宅的居住人群是万达广场最忠实拥趸的消费者；销售物业需要生活配套设施，万达广场的多种经营业态全方位满足了销售物业中不同人群的众多需求。各业态之间的良性循环不断地吸引着超高人气，商业氛围越来越浓，形成了独具特色的城市核心商圈，使销售物业的升值空间越来越大，创造了"万达广场就是城市中心"的口碑。

在运营过程中，销售物业为万达广场提供了稳定的人流保障；万达商业同时也为销售物业提供更具便捷的生活配套，为销售物业的升值、宜居生活创造了条件。

3. 设计角度

万达广场设计了一种全新的生活方式，规划设计整合了商业资源、景观资源、休闲娱乐资源和居住资源；各业态既相对独立又相互依存，高度和谐统一。规划设计不再简单的是一个地块的规划，而是一个"城"的概念，面积虽小，功能兼备，业态与业态之间的互动及依赖更体现出规划上的科学性及联动性。因销售物业的存在，万达广场规划称得上真正意义的规划。同时，销售物业解决了纯商业规划无法解决的问题：利用非主要干道侧规划销售物业，确保安静；利用主要道路规划商业，确保展示面的商业气氛，避免了非主要街道商业不"火"的问题。在此规划背景下，地块的各方向、各种相对复杂的周边环境均有良好的结合和利用。

在建筑设计中，万达大商业部分体量大，气势磅礴，

for Wanda Plaza while the latter advertises good reputation and supporting living facilities for the former in the development process. The two are mutually supplementary and interdependent, realizing optimized combination and positive cycle of product chain and capital chain (Figure 4).

2. OPERATION PERSPECTIVE

Popularity is necessary for operation of Wanda Plaza. Those working in the office buildings and living in the apartments and residences are loyal clients of Wanda Plaza. Supporting living facilities are necessary for the Properties for Sale and multiple business types in Wanda Plaza meets comprehensive demands of different groups in the Properties for Sale. The positive cycle among business types attracts more and more people and business atmosphere becomes stronger and stronger, forming a unique core business circle in the city. The Properties for Sale become more and more valuable, establishing the reputation of "Wanda Plaza is the city center".

In the operation process, the Properties for Sale provide stable stream of people for Wanda Plaza while Wanda commerce provides more convenient supporting living facilities for the Properties for Sale, creating conditions for its appreciation and comfortable life.

3. DESIGN PERSPECTIVE

Wanda Plazas are designed with a brand new life style. The planning design integrates commercial resources, landscape resources, leisure and entertainment resources and residential resources which are independent and interdependent in uniformed harmony. It is never about planning for a piece of land any more but the concept of "city". Small area contains complete functions that interaction and dependency among industries reflect scientific and systematic features. The existence of the Properties for Sale makes the planning of Wanda Plaza full of real significance. Meanwhile, the Properties for Sale resolve problems that pure commercial planning cannot: the Properties for Sale is

但没有制高点;写字楼部分会点亮城市视野的新鲜感,摩天大厦硬朗挺拔的外观设计,建立一个更加生动的商业形象和视觉体验。地标性的建筑设计,提高了整个项目品质;住宅部分凹凸的线脚、变幻的阴影,丰富了万达广场的建筑形态,使万达综合体具备了柔情的一面。销售物业使万达广场造型丰富、体块多样,不同角度均有品味、回味的内涵。

住宅部分园林式的居住环境,安逸而不失繁华设计的初衷,融入颇多人文与舒逸的元素,并将"舒享人居"作为生命的第一要义;商业部分环境热情、奔放、大气,兼备都市和商务的氛围。二者相辅相成,让参与其中者有一种"空间蒙太奇"的建筑体验,相比其他项目更具特点及吸引力,体验着"一半是火焰,一半是海水"的浓缩人生。

地下室利用居住、办公与购物人群使用时间不同,采用错峰停车位的布局设计,将人性化、车流动线考虑得丝丝入扣,充分利用了空间的使用效率,有效地节约社会资源和土地资源。在时间、空间的两维平台中,充分地释放生活资源的优势,并使之得到提升。

在设计中,销售物业不仅弥补了自持物业空白,也在复合物业中产生了聚合增值,丰富了万达广场的产品设计类型,使商业与非商业融为一体,相互补充,两者形成完美"解构"与"组合"。

销售物业因为存在于万达广场中而更具吸引力,更具卖点,更加完善;万达广场因销售物业的存在而更充实,更具彰显经典的标杆力量,为万达广场的精彩增添了浓墨重彩的一笔。

planned at minor trunk road for quite environment and business at main trunk road for display of commercial atmosphere, avoiding the problem of depression at minor trunk road. The planning brings good combination and use of land in all directions with complicated surrounding environment.

In the architectural design, the large business area of Wanda has big volume and great momentum without commanding height. The office building brings fresh feeling to city and skyscraper establishes a vivid business image and visual experience with its high and straight appearance. The skyscraper improves quality of the whole project. The convex architrave and changing shadow for the residential part enrich architectural shape of Wanda Plaza, showing gentle side of Wanda complex. The Properties for Sale enable Wanda Plaza in rich shapes and diversified places, making it full of taste and content in different angles.

The garden-type living environment is comfortable and prosperous, involving many humanity and comfort elements and taking "comfort, enjoyment and habitat" as the prior elements in the life. The commercial environment is full of enthusiasm and magnificence, having urban and commercial atmosphere. The living and commercial environment is supplementary, bringing "space montage" experience to people. Compared to other projects, Wanda Plaza has more attraction, making participators experience the colorful life.

As residential, official and shopping groups use the underground parking in different time periods of a day, the idea of peak shifting has been applied to fully consider humanity and flow and take full advantage of space, effectively saving the social resources and land. Advantages of the life resources are fully released and improved in aspects of time and space.

The Properties for Sale supplement holding properties and generate aggregated value in composite property in the design. Product types of Wanda Plaza are enriched to integrate the commercial part and non-commercial parts. The two parts are mutually supplementary to form perfect "deconstruction" and "combination".

The Properties for Sale are more attractive, easier for sales and perfect because of Wanda Plaza. Wanda Plaza becomes more powerful in display of classical building for existence of the Properties for Sale, which play an important role for wonderful Wanda Plaza.

WANDA PROPERTIES FOR SALE 2013
万达销售物业 2013

文／万达商业地产设计中心技术管理部副总经理　白雪松

对于万达商业地产股份有限公司来说，2013年是不平凡的一年。商业地产的销售额连续几年保持30%以上的增长率，收入1456亿元，同比增长32%；新开业万达广场18座，全国累计开业万达广场85座。

万达销售物业2013年的加速发展，对万达商业地产设计中心的人才专业结构和管理能力提出了更高的要求。2012年8月南北方项目管理中心设计部（设计中心的前身）扩编为90人，增加了结构和机电专业人员。南北方项目管理中心设计部迅速适应了新的工作职责，在2012年完成了《销售物业设计管控手册》、《设计管控计划流程及模板》、"销售物业上线模块化节点"和《销售物业定额设计》等制度的编制。2013年，销售物业设计管控制度不断完善，各级专业人员的能力素质日益提升，各项目产品品质得到了大幅度提升。南北方设计部全年对103个项目的销售物业进行全专业全过程管控，完成了6575项工作节点。35个售楼处、233套样板间、19个项目的营销样板段示范区高品质开放；14个项目销售物业产品取得了绿建设计一星认证，另有1个项目取得了绿建二星级认证。

2013年方案设计不断创新，推出了大连东港万达公馆、福建泉州万达公馆、长沙开福万达公馆、无锡宜兴万达公馆等高品质项目，总图、户型、立面、内装、景观全面突破，获得了市场的追捧，持续热销，其销售单价远高于同样地段的其他项目售价。沈阳奥体万达公馆、南宁青秀万达公馆、东莞厚街万达华府等35个售楼处精品迭出，233套样板间美轮美奂，19个项目样板段人流如织（图1、图2）。

It is not an usual year for Wanda Commercial Estate Co., Ltd. The sales revenue of the Commercial Properties has kept the growth rate of over 30% for several years with an income of 145.6 billion Yuan, up 32% year on year. 18 Wanda Plazas were opened in 2013, accumulating of a total of 85 Wanda Plazas in the whole country.

Rapid development of Wanda Properties for Sale in 2013 poses higher requirements for organization structure and management capability of the Design Center of Wanda Commercial Property. The Design Department of the South and North Project Management Center (predecessor of the Design Center) was expanded to 90 persons in August 2012 with an increasing of structural and mechanical professionals. The department rapidly adapted to new duties and completed formulation of *Design and Control Manual of Properties for Sale*, *Process and Sample of Design and Control Plan*, Modulation Milestones for Online of Properties for Sale and *Quota Design of Properties for Sale* in 2012. In 2013, the improved design and control systems of Properties for Sale and capabilities of professionals in each level lead to great quality improvement of products in each project. The north and south design department has controlled 103 projects of Properties for Sale during the whole process with the accomplishment of 6,575 project milestones. Besides, those projects were opened with high quality including 35 sales office, 233 prototype-rooms and 19 marketing model demonstration areas. 14 of them were awarded One-Star Green Building Design Label and one of them won Two-Star Green Building Operation Label.

（图1）福建泉州万达公馆建筑外立面

产品的高品质呈现，不仅需要方案创新，也得益于制度管控。第三方审图的应用、设计全程评审会制度、卖场日报制度、品质周报制度、定额设计、设计中心项目经理负责制、培训体系等一系列的管理体系日益完善，牢牢地控制住了项目的设计进度、品质、质量和成本。

（图2）烟台芝罘万达公馆售楼处景观

2013年完成了54项重要研发成果。《销售物业品质管控手册》的编制，为2014年品质管控大幅度提升将起到重要的作用；SOHO公寓全专业模块、售楼处模块、底商模块、样板间内装模块等模块的应用，以及外保温、砌筑材料、防水材料、外门窗、商铺美陈、内装景观现场管控标准等一系列设计标准的实施，为项目的安全和品质保驾护航。

2013年应用和继续完善了销售物业设计管控手册，实现了工作方法标准化；项目管理中心设计部、项目公司设计部全员业务培训和考试；过程中强化管控，设计中心每周检查；月考核、年汇总，让项目公司设计部找到工作目标，鞭策落后，奖励先进（图3）。

（图3）销售物业设计管控手册

Project designs in 2013 reflect persistent innovation and the launched high-quality projects are popular in the market for comprehensive breakthrough in the general drawing, house type, façade, interior decoration and landscape. The projects of Dalian Donggang Wanda Mansion, Fujian Quanzhou Wanda Mansion, Changsha Kaifu Wanda Mansion and Wuxi Yixing Wanda Mansion were sold well and its unit price was far higher than that of other projects in the same place. Quality products are launched in 35 sales offices including Shenyang Olympic Wanda Mansion, Nanning Qingxiu Wanda Mansion and Dongguan Houjie Wanda Palace. People are packed in 19 project prototype areas and 233 magnificent prototype rooms (Figure 1-2).

High quality presentation of products both require innovative plan and system control. A series of management systems were increasingly improved to better control design process, quality and cost of project including the application of the third party's check of drawings, review meeting system in the whole design process, daily report system of Sales Place, weekly report of quality system, quota design, project manager's responsibility system of the Design Center and training system, etc.

54 important R&D achievements were completed in 2013. Compilation of *Quality Control Manual of Properties for Sale* has greatly improved quality control in 2014. Module application of SOHO apartment full specialty module, sales office module, floor store module, interior decoration module in the prototype room and implementation of design standards including control standards of external insulation, masonry materials, anti-water material, external doors and windows, store display and landscape site of interior decoration, etc, provide guarantee for safety and quality of projects.

Design and Control Manual of Properties for Sale was applied and revised in 2013 to standardize the management methods. Business training and examination were conducted for members of the Design Department of Project Management Center and Project Companies. Control is reinforced in the process that the Design Center conducts weekly check, monthly examination and annual summary, making the Design Department of the Project Company identify working targets with the purpose of supervision (Figure 3).

2013年11月,南北方项目管理中心设计部与万达商业规划研究院统一划归商业规划设计系统序列,南北方项目管理中心设计部将合并为商业地产设计中心。规划系统经过整合联动,必将增强万达商业地产设计管控的综合实力。

2014年,设计中心的重点工作是创新。

项目管控实现方案创新,不断推出行业领先、引领市场需求的新产品,体现设计的价值。慧云智能化管理系统、室内空气环境净化措施等一系列重点研发成果将在2014年应用,获得市场检验。

The Design Department of the South and North Project Management Center and Wanda Commercial Planning & Research Institute are classified into commercial planning and design system in November 2013. The Design Department of the South and North Project Management Center is combined into Wanda Commercial Estate Design Center. Integration of planning system will strengthen comprehensive management skills of design and control of Wanda Commercial Properties.

The Design Center will focus on innovation in 2014.

Project management is a way to realizes innovative plan and launch new products that can lead the

(图4)2014版销售物业品质管控手册

管理不断创新，把标准文字变成标准图纸、图片。研发课题总体思路是策划前置，品质优先。2014版《管理制度汇编》、2014版《安全管理制度汇编》的编制，将指导今后的设计管控方向；《销售物业品质管控手册》的落地应用、《销售物业设计管控手册》的修订，提升项目设计的品质安全；百变户型、收纳空间等创新研发，引领万达产品升级；实现全程信息化管控，实现OA、计划模块与图文档系统三个信息化平台的对接互动，录入全部历史档案。（图4）

集团提出，无论企业发展到什么规模，"永远坚持品质提升"。商业地产设计中心也永远将品质提升作为工作重点，持之以恒地不懈努力，不断创新，不断走向成功。

光阴似箭，过去的2013年在万达商业地产设计中心的发展中具有特殊意义。为此，我们在本次出版的年鉴中，特别收录了2013年万达商业地产设计中心所有同事的照片及姓名。同时借此出版图书之际，向那些曾经一同为万达商业地产设计中心工作、为万达的事业辛勤付出的同事们表示诚挚的敬意，并对所有为万达工作过的同仁及关注和支持万达的朋友们表示由衷的感谢！

trend in industry and meet market demand. This is a reflection of the design value. A series of key R&D achievements will be applied in 2014 including Huiyun Intelligent Management System and indoor air environment cleaning measures for market test.

Management should always be improved like reflecting standard words into more vivid drawings and pictures. The general thought of R&D project is advanced planning and quality priority. Compilation of *Management System Collection Version 2014* and *Security Management System Collection Version 2014* will be guidance of design and control from now on. Application of *Quality Control Manual of Properties for Sale* and revision of *Design and Control Manual of Properties for Sale* promote quality security of project design. Innovative changeable house types and storage function will update Wanda products. Information technology control is realized in the whole process. Connection and interaction among OA system, planning module and drawings and documents systems can assist Wanda in recording all business activities and making it history (Figure 4).

The Group puts forwards the principle of "Quality Improvement Forever" no matter how big the enterprise will become. The Design Center of the Commercial Properties will always take quality improvement as key work and make unremitting efforts to success one after another.

Time flies. The past 2013 has special significance in the development of the Design Center of Wanda Commercial Properties. And photos and names of colleagues in the Center are particularly collected into this published yearbook. At the same time, taking the opportunity of publishing this book, we would like to express our sincere gratitude and appreciation for colleagues and friends, especially those who have been working in the Design Center and those who made huge contribution for Wanda development.

POSTSCRIPT
后记

文／万达商业地产副总裁　赖建燕
（原万达商业地产高级总裁助理
兼商业规划研究院院长）

万达集团成立于1988年，自成立之初就从事住宅等销售类物业的开发，至今已走过26个年头，取得了辉煌的成就，在行业内创造了多个"第一"。1992年首届"质量万里行"活动中，万达民政街小区获得全国唯一一块"优质住宅工程"奖牌，同年万达走出大连到广州发展，成为中国第一家跨区域发展的房地产企业；2000年在建设部于人民大会堂召开的全国住宅优质工程表彰大会上，时任建设部部长俞正声对万达高度赞扬，亲自为王健林董事长颁奖。这是建设部成立以来树立的唯一房地产企业典型；2003年昆明滇池卫城项目成为全国第一个做环境评估的住宅小区。

万达在进行地产开发时尤其注重社会责任，大连雍景台项目是全国最早的节能住宅之一，之后万达在销售物业的开发设计中把绿色节能放在首要位置。自2011年开始，全部住宅项目均要求达到绿建星级标准。截至2013年底，万达销售物业共取得绿建星级项目39个，其中一项达到绿建二星标准，在绿色节能领域起到了模范带头作用。

万达在销售物业领域取得的辉煌成就有力地促进了企业的快速发展，至今已连续八年销售额每年以30%的速度增长。成为万达商业地产最基本的支柱业态，为万达集团可持续发展与转型做出了卓越的

Wanda Group was founded in 1988 as a company engaged in development of residential and other properties for sale. It has a history of 26 years and attained wonderful achievements. Wanda Group has been awarded many "No.1s" in the industry. In the first "China Quality Long March" activity in 1992, Minzheng Street Community of Wanda was awarded the only national medal of "Quality Residential Project" and in the same year, Wanda went out Dalian to develop in Guangzhou, becoming the first real estate company for cross-regional development in China. In the commendation conference of national quality residential projects held by Chinese Ministry of Construction (at present the Ministry of Housing and Urban-Rural Development) in the Great Hall of the People in 2000, Yu Zhengsheng, Minister of Construction at that time, highly praised Wanda Group and gave prize to Chairman Wang Jianlin in person. Wanda Group has been the only real estate model set up since establishment of the Ministry of Construction. The Dianchi Weicheng Project in Kunming becomes the first residential community for environmental assessment in China in 2003.

Wanda Group pays special attention to social responsibility in property development. The Yongjingtai Project in Dalian is one of the earliest energy-saving residential buildings in China and after that Wanda places green and energy saving at its priority in development and design of the Properties for Sale.

（图2）大连东港万达公馆精装

（图3）荆门万达华府售楼处夜景

（图1）天津海河东路商业街建筑夜景

(图4)长沙开福万达公馆售楼处景观

贡献。"明珠"、"花园"到"公馆"、"华府"等一系列高端销售类产品涌现,开创了万达特有的销售类物业产品体系。销售物业在多年的开发设计中得到了王健林董事长、丁本锡总裁、齐界总裁及集团各级领导的大力支持,形成从建筑、景观、精装、机电、结构、绿建等专业系统化的管理制度和技术标准,并纳入集团信息化管控体系当中,成为万达综合体模式不可或缺的重要组成部分(图1~图4)。

本《年鉴》的出版发行得益于所有参与其中的各级领导、各位同仁、合作供方、合作部门、出版发行机构等的大力支持,《年鉴》作品征集得到了各项目公司的大力支持。谨此向以上相关人员表示衷心的感谢。

万达商业地产设计中心成立于2014年4月,是由原南、北、文旅项目管理中心设计部合并重组而来。设计中心集中回顾了2013年及之前万达销售物业的优秀项目,编制本《年鉴》。这是万达销售物业的第一本图目式年鉴,在编辑2013年当年优秀销售物业项目的同时,也整理回顾了2013年之前的销售物业优秀项目,以后每年我们都将整理回顾当年的销售物业优秀作品,形成系列化的年鉴书刊,与业内、外人士广泛交流与分享。

Since 2011, all residential projects have been required to reach the star green building design label. Up to the end of 2013, 39 projects of Wanda Properties for Sale have been awarded star green building design labels with one up Two-Star Green Building Design Label, being a good example in the green and energy saving field.

Brilliant achievements of Wanda in the Properties for Sale greatly promote rapid development of the enterprise and the sales revenue has increased at a speed of 30% for continuous eight years. The Properties for Sale have become the basic pillar industry of Wanda Commercial Properties, making excellent contribution to sustainable development and transition of Wanda Group. A series of high-end sales products appear from "pearl", "garden" to "mansion", "palace", creating Wanda product system specific to the Properties for Sale. Chairman Wang Jianlin, President Ding Benxi, President Qi Jie and group leaders in all levels have intensively engaged in the development of the Properties for Sale for many years. The professional and systematic management system and technical standard are established including building, landscape, fine decoration, mechanical, structure, green building, etc, which are contained in Group's information technology control system and became an indispensable part of Wanda complex mode (Figure 1-4).

Publication of this Yearbook has been greatly supported by participants of leaders in each level, colleagues, cooperative supply parties, cooperative sectors and publication organization. Also, this Yearbook received much support from project companies in collecting of works. Sincere gratitude extended to the above persons.

Set up in April 2014, the Wanda Commercial Estate Design Center is a merger and reorganization of the original Design Departments of the South, North Project Management Centers as well as the Project Management Center of Wanda Cultural Tourism Planning & Research Institute. The Design Center compiles the Yearbook to review excellent projects in 2013 and the time before. Being the first graphic yearbook of Wanda Properties for Sale, the Yearbook introduces excellent projects of Wanda Properties for Sale in 2013, and also reviews excellent projects of the type before 2013. In the future, we will continue to compile and review excellent works of Wanda Properties for Sale of each year to establish series of yearbooks for sharing with friends inside and outside the industry.

PART G
WANDA COMMERCIAL PLANNING
PROJECT INDEX
项目索引

PROJECT INDEX
项目索引

ADMISSION PROJECTS
入伙项目

DALIAN DONGGANG WANDA MANSION
大连东港万达公馆
2010 / 10

软装设计单位　　　北京班迪斯室内设计有限责任公司

参与人员
精装：昌燕

FUJIAN QUANZHOU WANDA MANSION
福建泉州万达公馆
2013 / 05

建筑设计单位　　　深圳市清华苑建筑设计有限公司
外幕墙设计单位　　北京市金星卓宏幕墙工程有限公司
内装设计单位　　　上海茂迪建筑设计事务所
　　　　　　　　　上海文翰建筑装饰工程设计有限公司
　　　　　　　　　北京紫香舸国际装饰艺术顾问有限公司
景观设计单位　　　北京中建建筑设计院有限公司上海分公司
夜景照明设计单位　栋梁国际照明设计（北京）中心有限公司

参与人员
建筑：赵龙　　景观：武春雨　　精装：李金桥　　结构：吴铮
设备：董丽梅　　强电：刘国祥　　弱电：陈涛

CHANGSHA KAIFU WANDA MANSION
长沙开福万达公馆
2013 / 11

建筑设计单位　　　深圳华森建筑与工程设计顾问有限公司
　　　　　　　　　深圳华森建筑与工程设计顾问有限公司
内装设计单位　　　上海品仓建筑室内设计有限公司
景观设计单位　　　深圳市致道思维建筑景观有限公司
夜景照明设计单位　大连路明光电工程有限公司

参与人员
建筑：何伟华　　景观：郑锐　　精装：李春阳　　机电：桑国安　　结构：邵强

WUHAN JIYUQIAO WANDA MANSION
武汉积玉桥万达公馆
2013 / 10

建筑设计单位　　　上海联创建筑设计有限公司
外幕墙设计单位　　北京市金星卓宏幕墙工程有限公司
内装设计单位　　　上海高迪建筑工程设计有限公司
景观设计单位　　　上海帕莱登建筑景观咨询有限公司
夜景照明设计单位　北京市建筑设计研究院

参与人员
建筑：何伟华　　景观：杨辉　　精装：李春阳　　机电：赵前卫
结构：杜刘生

JINAN WEIJIAZHUANG WANDA MANSION
济南魏家庄万达公馆
2010 / 09

建筑设计单位　　　上海联创建筑设计有限公司
内装设计单位　　　上海高迪建筑工程设计有限公司
景观设计单位　　　上海联创建筑设计有限公司

参与人员
建筑：王福魁　　景观：李琰　　精装：昌燕　　机电：朱应勇　　结构：李鹏

SHIJIAZHUANG YUHUA WANDA MANSION
石家庄裕华万达公馆
2013 / 04

建筑设计单位　　　北京建筑设计研究院
内装设计单位　　　上海发现建筑装饰设计工程有限公司
景观设计单位　　　致道国际设计顾问（香港）有限公司
　　　　　　　　　中国建筑设计研究院

参与人员
建筑：叶啸　　景观：陈霎明　　精装：昌燕　　机电：冯俊
结构：邵强

ZHANGZHOU BIHU WANDA PALACE
漳州碧湖万达华府
2012 / 12

建筑设计单位　　　万达商业规划研究院有限公司
　　　　　　　　　厦门中福元建筑设计研究院
外幕墙设计单位　　华凯国际派特建筑设计（北京）有限公司
内装设计单位　　　深圳市华辉装饰工程有限公司
　　　　　　　　　北京市建筑装饰设计院有限公司
景观设计单位　　　上海现代建筑设计（集团）有限公司
　　　　　　　　　深圳市致道思维建筑景观设计有限公司
夜景照明设计单位　深圳市千百辉照明工程有限公司

参与人员
建筑：赵龙　　景观：薛奇　　结构：吴铮　　设备：薛勇　　强电：刘国祥　　弱电：陈涛

ZHENGZHOU ERQI WANDA PALACE
郑州二七万达华府
2013 / 07

建筑设计单位　　　北京中联环建文建筑设计有限公司
　　　　　　　　　北京五合国际建筑设计咨询有限公司
　　　　　　　　　中国建筑科学研究院
内装设计单位　　　河南砺石装饰设计工程有限公司
景观设计单位　　　上海帕莱登建筑景观咨询有限公司

参与人员
建筑：马林峰　　景观：张金菊　　精装：崔宇航　　机电：冯俊　　结构：邵强

FUJIAN PUTIAN WANDA PALACE
福建莆田万达华府
2013 / 09

建筑设计单位　　　深圳市清华苑建筑设计有限公司
外幕墙设计单位　　上海旭密林幕墙有限公司
内装设计单位　　　腾远设计事务所有限公司
景观设计单位　　　北京中建建筑设计院有限公司上海分公司
夜景照明设计单位　深圳市千百辉照明工程有限公司

参与人员
建筑：靳松　　景观：黄建好　　精装：李金桥　　结构：吴铮
设备：薛勇　　强电：宋波　　弱电：陈涛

SALES PLACES
销售卖场

SHENYANG OLYMPIC WANDA MANSION: SALES OFFICE/ PROTOTYPE ROOM
沈阳奥体万达公馆－售楼处/样板间
2012 / 05

建筑设计单位	大连市建筑设计研究院有限公司
外幕墙设计单位	中国建筑科学研究院
内装设计单位	广州思则装饰设计有限公司
景观设计单位	笛东联合（北京）规划设计顾问有限公司
夜景照明设计单位	北京三色石环境艺术设计有限公司

参与人员
建筑：曾静　　景观：郑锐　　精装：秦峥

NANNING QINGXIU WANDA MANSION: SALES OFFICE/ PROTOTYPE ROOM
南宁青秀万达公馆－售楼处/样板间
2013 / 12

建筑设计单位	上海联创建筑设计有限公司
外幕墙设计单位	北京金星卓宏幕墙工程有限公司
内装设计单位	北京米多芬建筑设计咨询有限公司
	广州思哲装饰设计有限公司
景观设计单位	上海赛特康斯景观设计咨询有限公司(SCI)
夜景照明设计单位	深圳市金达照明股份有限公司

参与人员
建筑：靳松　　景观：黄建好　　精装：李金桥　　结构：杨募
设备：霍雪影　　强电：宋波　　弱电：陈涛

CHANGSHA KAIFU WANDA MANSION: SALES OFFICE/ PROTOTYPE ROOM
长沙开福万达公馆－售楼处/样板间
2013 / 11

建筑设计单位	KSP
外幕墙设计单位	北京市金星卓宏幕墙工程有限公司
内装设计单位	广州戴维建筑设计工程有限公司
景观设计单位	深圳市致道景观有限公司
夜景照明设计单位	北京市建筑设计研究院

参与人员
建筑：王浩　　景观：郑锐　　精装：秦峥

DONGGUAN DONGCHENG WANDA MANSION: SALES OFFICE/ PROTOTYPE ROOM
东莞东城万达公馆－售楼处/样板间
2012 / 08

建筑设计单位	深圳市清华苑建筑设计有限公司
外幕墙设计单位	厦门开联装饰工程有限公司
内装设计单位	上海茂迪建筑设计事务所
	深圳安星装饰设计工程有限公司
景观设计单位	广东华方工程设计有限公司
夜景照明设计单位	北京三色石环境艺术设计有限公司

参与人员
建筑：杜文天　　景观：邹昊　　精装：朱卓毅　　结构：杨募
设备：董丽梅　　强电：关向东　　弱电：陈涛

YANTAI ZHIFU WANDA MANSION: SALES OFFICE/ PROTOTYPE ROOM
烟台芝罘万达公馆－售楼处/样板间
2013 / 07

建筑设计单位	上海联创建筑设计有限公司
外幕墙设计单位	深圳蓝波绿建集团股份有限公司
内装设计单位	北京米多芬建筑设计咨询有限公司
	上海联创建筑设计有限公司
夜景照明设计单位	北京鱼禾光环境有限公司

参与人员
建筑：陈文娜　　景观：郑锐　　精装：钟山

NANJING JIANGNING WANDA MANSION: SALES OFFICE
南京江宁万达公馆－售楼处
2012 / 04

内装设计单位	广州思则装饰设计有限公司
景观设计单位	江苏天正建筑景观规划设计有限公司

参与人员
内装：陈晖　　景观：顾东方

WUHAN K4 WANDA MANSION: SALES OFFICE/ PROTOTYPE ROOM
武汉K4万达公馆－售楼处/样板间
2013 / 04

建筑设计单位	上海联创建筑设计有限公司
外幕墙设计单位	北京市金星卓宏幕墙工程有限公司
内装设计单位	法国米多芬建筑设计师事务所
景观设计单位：	上海兴田建筑工程设计事务所

参与人员
建筑：何伟华　　景观：杨辉　　精装：李春阳

XI'AN DAMINGGONG WANDA MANSION: SALES OFFICE/ PROTOTYPE ROOM
西安大明宫万达公馆－售楼处/样板间
2012 / 04

建筑设计单位	北京中联环建文建筑设计有限公司
外幕墙设计单位	珠海兴业绿色建筑科技有限公司
内装设计单位	北京紫香舸国际装饰艺术顾问有限公司
夜景照明设计单位	陕西大地重光景观照明设计工程有限公司

参与人员
建筑：孙静　　精装：崔宇航

ZHENGZHOU JINSHUI WANDA MANSION: SALES OFFICE
郑州金水万达公馆－售楼处
2013 / 12

建筑设计单位	豪斯泰勒张思图德建筑设计咨询（上海）有限公司
	华凯派特建筑设计（北京）有限公司
外幕墙设计单位	北京市建筑设计研究院有限公司
内装设计单位	上海孙李建筑设计咨询有限公司
	美伦美（北京）国际贸易有限公司
	北京班迪斯室内设计有限公司
景观设计单位	上海兴田建筑工程设计事务所

参与人员
建筑：董莉　　景观：张金菊　　精装：杨磊　　机电：冯俊　　结构：邵强

DONGGUAN HOUJIE WANDA PALACE: SALES OFFICE

东莞厚街万达华府 -
售楼处

2013 / 12

建筑设计单位　　广东华方工程设计有限公司
外幕墙设计单位　　北京和平幕墙工程有限公司
内装设计单位　　上海蓝天房屋装饰工程有限公司
景观设计单位　　深圳威瑟本景观设计有限公司
夜景照明设计单位　广东华方工程设计有限公司

参与人员
建筑：杜文天　　景观：邹昊　　精装：胡延峰　　结构：杨募
设备：董丽梅　　强电：关向东　　弱电：陈涛

DEZHOU WANDA PALACE: SALES OFFICE/PROTOTYPE ROOM

德州万达华府 -
售楼处 / 样板间

2013 / 09

建筑设计单位　　北京莫克建筑规划设计咨询有限公司
外幕墙设计单位　　北京市金星卓宏幕墙工程有限公司
内装设计单位　　北京迪文建筑装饰设计有限公司
　　　　　　　　南京嘉朗都市环境工程有限公司
景观设计单位　　宝佳丰（北京）国际建筑景观规划设计有限公司
夜景照明设计单位　北京鱼禾光环境设计有限公司

参与人员
建筑：赵立群　　景观：范志满　　精装：陈玉斌　　机电：桑国安
结构：邵强

ZHEJIANG JIAXING WANDA PALACE: PROTOTYPE ROOM

浙江嘉兴万达华府 -
样板间

2013 / 11

内装设计单位　　青岛腾远设计事务所有限公司
景观设计单位：　浙江宏正建筑设计有限公司

参与人员
建筑：周昳晗　　精装：杨琼　　结构：吴铮　　设备：霍雪影
强电：刘国祥　　弱电：陈涛

JIAMUSI WANDA PALACE: SALES OFFICE

佳木斯万达华府 -
售楼处

2013 / 07

建筑设计单位　　中旭建筑设计有限责任公司
外幕墙设计单位　　北京建黎铝门窗幕墙有限公司
内装设计单位　　青岛腾远设计事务所有限公司
夜景照明设计单位　蒙尔赛照明技术集团有限公司

参与人员
建筑：赵宁宁　　精装：钟山

MIANYANG CBD WANDA PALACE: PROTOTYPE ROOM

绵阳CBD万达华府 -
样板间

2013 / 04

内装设计单位　　广州思哲装饰设计有限公司

参与人员
建筑：赵龙　　精装：宋之煜

WEINAN WANDA PALACE: SALES OFFICE/PROTOTYPE ROOM

渭南万达华府 -
售楼处 / 样板间

2013 / 08

建筑设计单位　　西安建筑科技大学设计研究院
外幕墙设计单位　　西安建筑科技大学设计研究院
内装设计单位　　北京中联环建文建筑设计有限公司
　　　　　　　　青岛腾远设计事务所有限公司
景观设计单位　　西安西野景观规划设计有限公司
夜景照明设计单位　北京三色石环境艺术设计有限公司

参与人员
建筑：叶啸　　景观：郑锐　　精装：刘洋　　机电：罗粤峰　　结构：杜刘生

FOSHAN NANHAI WANDA PALACE: PROTOTYPE ROOM

佛山南海万达华府 -
样板间

2013 / 09

内装设计单位　　广州思哲装饰设计有限公司

参与人员
建筑：曹鹏　　精装：胡延峰

JINING WANDA PALACE: PROTOTYPE ROOM

济宁万达华府 - 样板间

2013 / 09

内装设计单位　　青岛腾远设计事务所有限公司

参与人员
精装：钟山

WENZHOU PINGYANG WANDA PALACE: PROTOTYPE ROOM

温州平阳万达华府 -
样板间

2013 / 08

内装设计单位　　腾远设计事务所有限公司

参与人员
建筑：曹鹏　　精装：单萍

CHONGQING BA'NAN WANDA PALACE: SALES OFFICE
重庆巴南万达华府－售楼处
2013 / 11

建筑设计单位	机械工业第三设计研究院
外幕墙设计单位	广东省华城建筑设计有限公司
内装设计单位	腾远设计事务所有限公司
景观设计单位	重庆九禾园林景观设计工程有限公司
夜景照明设计单位	机械工业第三设计研究院

参与人员

建筑: 靳松　　景观: 薛奇　　精装: 胡延峰　　结构: 黄达志
设备: 霍雪影　　强电: 宋波　　弱电: 陈涛

DEMONSTRATION AREAS
示范区

DEMONSTRATION AREA OF DALIAN HIGH-TECH ZONE WANDA MANSION
大连高新万达公馆示范区
2013 / 08

建筑设计单位	广东启源建筑工程设计有限公司众成分公司
外幕墙设计单位	北京市金星卓宏幕墙工程有限公司
内装设计单位	北京紫香舸国际装饰艺术顾问有限公司
景观设计单位	深圳市致道思维景设计有限公司
夜景照明设计单位	深圳市千百辉照明工程有限公司

参与人员

建筑: 叶啸　　景观: 张金菊　　精装: 李春阳　　机电: 赵前卫　　结构: 李鹏

DEMONSTRATION AREA OF DONGGUAN DONGCHENG WANDA MANSION
东莞东城万达公馆示范区
2013 / 07

建筑设计单位	深圳市清华苑建筑设计有限公司
外幕墙设计单位	厦门开联装饰工程有限公司
内装设计单位	广州韦格斯杨设计有限公司
	广州市城市组设计有限公司
景观设计单位	宝佳丰(北京)国际建筑景观规划设计有限公司
夜景照明设计单位	北京三色石环境艺术设计有限公司

参与人员

建筑: 杜文天　　景观: 黄建好　　精装: 朱卓毅　　结构: 杨募
设备: 董丽梅　　强电: 关向东　　弱电: 陈涛

DEMONSTRATION AREA OF TIANJIN HAIHEDONGLU COMMERCIAL STREET
天津海河东路商业街示范区
2013 / 08

建筑设计单位	北京市建筑设计研究院
外幕墙设计单位	深圳蓝波幕墙及光伏工程有限公司
景观设计单位	棕榈园林股份有限公司
夜景照明设计单位	北京三色石环境艺术设计有限公司

参与人员

建筑: 孙静　　景观: 范志满

DEMONSTRATION AREA OF NANNING QINGXIU WANDA MANSION
南宁青秀万达公馆示范区
2013 / 12

内装设计单位	北京米多芬建筑设计咨询有限公司
	广州思则装饰设计有限公司
景观设计单位	上海赛特康斯景观设计咨询有限公司（SCI）
夜景照明设计单位	深圳市金达照明股份有限公司

参与人员
建筑: 靳松　景观: 黄建好　精装: 李金桥　结构: 杨募
设备: 霍雪影　强电: 宋波　弱电: 陈涛

DEMONSTRATION AREA OF WUHAN K9 WANDA MANSION
武汉 K9 万达公馆示范区
2013 / 07

内装设计单位	北京紫香舸国际装饰艺术顾问有限公司
景观设计单位	广州棕榈园林股份有限公司
夜景照明设计单位	栋梁国际照明设计（北京）中心有限公司

参与人员
景观: 杨辉　精装: 李春阳　机电: 冯俊　结构: 杜刘生

DEMONSTRATION AREA OF WUXI YIXING WANDA MANSION
无锡宜兴万达公馆示范区
2013 / 05

建筑设计单位	上海中星志成建筑设计有限公司
外幕墙设计单位	北京市金星卓宏幕墙工程有限公司
内装设计单位	北京天秤座装饰设计有限公司
景观设计单位	中国建筑设计研究院
夜景照明设计单位	北京三色石环境艺术设计有限公司

参与人员
建筑: 李晅荣　景观: 顾东方　精装: 宋之煜　结构: 黄达志
设备: 薛勇　强电: 刘国祥　弱电: 陈涛

DEMONSTRATION AREA OF XI'AN DAMINGGONG WANDA MANSION
西安大明宫万达公馆示范区
2012 / 04

景观设计单位	棕榈园林股份有限公司

参与人员
建筑: 孙静　景观: 杨辉　机电: 朱应勇　结构: 杜刘生

DEMONSTRATION AREA OF MA'ANSHAN WANDA PALACE
马鞍山万达华府示范区
2013 / 10

建筑设计单位	上海中星志成建筑设计有限公司
	上海思亚建筑设计咨询有限公司
外幕墙设计单位	北京市金星卓宏幕墙工程有限公司
	上海旭密林幕墙有限公司
内装设计单位	青岛腾远设计事务所有限公司
景观设计单位	北京易兰建筑规划设计有限公司
夜景照明设计单位	深圳普莱思照明设计顾问有限责任公司

参与人员
建筑: 李晅荣　景观: 武春雨　精装: 李万顺　结构: 黄达志
设备: 董丽梅　强电: 宋波　弱电: 陈涛

DEMONSTRATION AREA OF QINGDAO LICANG WANDA PALACE COMMERCIAL STREET
青岛李沧万达华府商业街示范区
2013 / 09

建筑设计单位	青岛腾远建筑设计事务所有限公司
外幕墙设计单位	北京金星卓宏幕墙工程有限公司
景观设计单位	上海赛特斯康景观设计咨询有限公司
夜景照明设计单位	青岛汇海泉景观工程有限公司
美陈设计单位	青岛汇海泉景观工程有限公司

参与人员
建筑: 叶啸　景观: 张金菊、郑锐

DEMONSTRATION AREA OF DONGGUAN CHANG'AN WANDA PALACE
东莞长安万达华府示范区
2013 / 06

建筑设计单位	清华苑建筑设计
内装设计单位	北京米多芬建筑设计咨询有限公司
景观设计单位	北京中建建筑设计院有限公司上海分公司
夜景照明设计单位	北京市建筑设计研究院 BIAD 灯光工作室

参与人员
建筑: 杜文天　景观: 乔勇　精装: 胡延峰　结构: 杨募
设备: 董丽梅　强电: 关向东　弱电: 陈涛

DEMONSTRATION AREA OF FOSHAN NANHAI WANDA PALACE
佛山南海万达华府示范区
2013 / 08

建筑设计单位	广州思哲设计院有限公司
外幕墙设计单位	深圳蓝波幕墙及光伏工程有限公司
内装设计单位	广州思哲设计院有限公司
景观设计单位	上海帕莱登建筑景观咨询有限公司
夜景照明设计单位	北京万德门特城市照明有限公司

参与人员
建筑: 曹鹏　景观: 薛奇　精装: 胡延峰　结构: 杨募
设备: 董丽梅　强电: 关向东　弱电: 陈涛

DEMONSTRATION AREA OF CHANGZHOU WUJIN WANDA PALACE
常州武进万达华府示范区
2013 / 12

建筑设计单位	上海兴田建筑设计事务所
外幕墙设计单位	北京市金星卓宏幕墙工程有限公司
内装设计单位	青岛腾远设计事务所有限公司
景观设计单位	上海兴田建筑设计事务所
夜景照明设计单位	北京三色石环境艺术设计有限公司

参与人员
建筑: 牛辉哲　景观: 武春雨　精装: 宋之煜　结构: 黄达志
设备: 董丽梅　强电: 关向东　弱电: 陈涛

DEMONSTRATION AREA OF WEIFANG WANDA PALACE
潍坊万达华府示范区
2013 / 09

内装设计单位	山东福缘来装饰有限公司
夜景照明设计单位	北京三色石环境艺术设计有限公司

参与人员

建筑: 辛小林　　精装: 孙洁

DEMONSTRATION AREA OF ANHUI BENGBU WANDA PALACE COMMERCIAL STREET
安徽蚌埠万达华府商业街示范区
2013 / 04

建筑设计单位	中建国际（深圳）建筑设计顾问有限公司
	上海兴田建筑工程设计事务所
外幕墙设计单位	北京市金星卓宏幕墙工程有限公司
景观设计单位	上海兴田建筑工程设计事务所
夜景照明设计单位	北京三色石环境艺术设计有限公司

参与人员

建筑: 李军　　景观: 薛奇　　结构: 黄达志
设备: 霍雪影　　强电: 宋波　　弱电: 陈涛

WANDA CHINESE-STYLE COURTYARD
万达中式四合院
2012 / 10

建筑设计单位	中旭建筑设计有限责任公司
内装设计单位	中旭建筑设计有限责任公司
景观设计单位	中旭建筑设计有限责任公司

参与人员

建筑: 王福魁　　景观: 郑锐　　精装: 钟山

2013
万达商业规划
销售类物业

WANDA COMMERCIAL
PLANNING 2013
PROPERTIES FOR SALE

尹强　刘拥军　林树郁　王福魁　霍雪影
昌燕　杨琼　董丽梅　顾东方　李金桥
朱卓毅　郑锐　杨威　杜文天　曾静
黄建好　邹昊　纪文青　李春阳　胡延峰
李军　赵立群　李晅荣　张金菊　钟山
靳松　雷宇　叶啸　白雪松　陈文娜
文善平　牛辉哲　吴铮　关向东　刘国祥
赵前卫　陈海燕　付东滨　刘征　黄达志
宋波　董莉　宋之煜　陈涛　杜刘生
邵强　李琰　刘大伟　武春雨　赵龙
曹鹏　薛勇　朱应勇　冯俊　桑国安
高伟杰　李万顺　孙志超　郑德廷　单萍
张东光　李卓东　莫疆　王凡　俞小华
范志满　李鹏　张克　张爱珍　周昳晗
赵宁宁　薛奇　杨磊　刘洋　黄文城
周恒　栾赫

图书在版编目（CIP）数据

万达商业规划 2013：销售类物业 / 万达商业地产设计中心主编.
-- 北京：中国建筑工业出版社，2014.12
ISBN 978-7-112-17558-1

Ⅰ.①万… Ⅱ.①万… Ⅲ.①商业区—城市规划—中国 Ⅳ.①TU984.13

中国版本图书馆CIP数据核字 (2014) 第 277958 号

责任编辑：徐晓飞　张　明
执行编辑：郑　锐
美术编辑：康　宇　陈　唯
英文翻译：喻蓉霞　王晓卉　郝　婧
责任校对：姜小莲

万达商业规划 2013：销售类物业
万达商业地产设计中心　主编

*

中国建筑工业出版社出版、发行（北京西郊百万庄）
各地新华书店、建筑书店经销
北京雅昌艺术印刷有限公司制版
北京雅昌艺术印刷有限公司印刷

*

开本：787×1092毫米　1/8　印张：38　字数：760千字
2014年12月第一版　2014年12月第一次印刷
定价：800.00元
ISBN 978-7-112-17558-1
(26747)

版权所有　翻印必究
如有印装质量问题，可寄本社退换
（邮政编码100037）